The last SHELBY COBRA

My times with Carroll Shelby

From Veloce Publishing:

Biographies
A Chequered Life – Graham Warner and the Chequered Flag (Hesletine)
A Life Awheel – The 'auto' biography of W de Forte (Skelton)
Amédée Gordini ... a true racing legend (Smith)
André Lefebvre, and the cars he created at Voisin and Citroën (Beck)
Bunty – Remembering a gentleman of noble Scottish-Irish descent (Schrader)
Chris Carter at Large – Stories from a lifetime in motorcycle racing (Carter & Skelton)
Cliff Allison, The Official Biography of – From the Fells to Ferrari (Gauld)
Edward Turner – The Man Behind the Motorcycles (Clew)
Driven by Desire – The Desiré Wilson Story
First Principles – The Official Biography of Keith Duckworth (Burr)
Inspired to Design – F1 cars, Indycars & racing tyres: the autobiography of Nigel Bennett (Bennett)
Jack Sears, The Official Biography of – Gentleman Jack (Gauld)
Jim Redman – 6 Times World Motorcycle Champion: The Autobiography (Redman)
John Chatham – 'Mr Big Healey' – The Official Biography (Burr)
The Lee Noble Story (Wilkins)
Mason's Motoring Mayhem – Tony Mason's hectic life in motorsport and television (Mason)
Raymond Mays' Magnificent Obsession (Apps)
Pat Moss Carlsson Story, The – Harnessing Horsepower (Turner)
'Sox' – Gary Hocking – the forgotten World Motorcycle Champion (Hughes)
Tony Robinson – The biography of a race mechanic (Wagstaff)
Virgil Exner – Visioneer: The Official Biography of Virgil M Exner Designer Extraordinaire (Grist)

General
An Incredible Journey (Falls & Reisch)
Anatomy of the Classic Mini (Huthert & Ely)
Anatomy of the Works Minis (Moylan)
Armstrong-Siddeley (Smith)
Art Deco and British Car Design (Down)
Autodrome (Collins & Ireland)
Automotive A-Z, Lane's Dictionary of Automotive Terms (Lane)
Automotive Mascots (Kay & Springate)
Bahamas Speed Weeks, The (O'Neil)
Chevrolet Corvette (Starkey)
Chrysler 300 – America's Most Powerful Car 2nd Edition (Ackerson)
Chrysler PT Cruiser (Ackerson)
Cobra – The Real Thing! (Legate)
Competition Car Aerodynamics 3rd Edition (McBeath)
Competition Car Composites A Practical Handbook (Revised 2nd Edition) (McBeath)
Concept Cars, How to illustrate and design – New 2nd Edition (Dewey)
Cortina – Ford's Bestseller (Robson)
Dodge Challenger & Plymouth Barracuda (Grist)
Dodge Charger – Enduring Thunder (Ackerson)
Dodge Dynamite! (Grist)
Draw & Paint Cars – How to (Gardiner)
Drive on the Wild Side, A – 20 Extreme Driving Adventures From Around the World (Weaver)
Fate of the Sleeping Beauties, The (op de Weegh/Hottendorff/op de Weegh)

Ford Cleveland 335-Series V-8 engine 1970 to 1982 – The Essential Source Book (Hammill)
Ford F100/F150 Pick-up 1948-1996 (Ackerson)
Ford F150 Pick-up 1997-2005 (Ackerson)
Ford Focus WRC (Robson)
Ford GT – Then, and Now (Streather)
Ford GT40 (Legate)
Ford Midsize Muscle – Fairlane, Torino & Ranchero (Cranswick)
Ford Mustang II & Pinto 1970 to 80 (Cranswick)
Ford Small Block V-8 Racing Engines 1962-1970 – The Essential Source Book (Hammill)
Ford Thunderbird From 1954, The Book of the (Long)
India - The Shimmering Dream (Reisch/Falls (translator))
Inside the Rolls-Royce & Bentley Styling Department – 1971 to 2001 (Hull)
Intermeccanica – The Story of the Prancing Bull (McCredie & Reisner)
Le Mans Panoramic (Ireland)
Motor Racing – Reflections of a Lost Era (Carter)
Motor Racing – The Pursuit of Victory 1930-1962 (Carter)
Motor Racing – The Pursuit of Victory 1963-1972 (Wyatt/Sears)
Motor Racing Heroes – The Stories of 100 Greats (Newman)
Motorsport In colour, 1950s (Wainwright)
MV Agusta Fours, The book of the classic (Falloon)
N.A.R.T. – A concise history of the North American Racing Team 1957 to 1983 (O'Neil)
Northeast American Sports Car Races 1950-1959 (O'Neil)
Nothing Runs – Misadventures in the Classic, Collectable & Exotic Car Biz (Slutsky)
Racing Colours – Motor Racing Compositions 1908-2009 (Newman)
Schlumpf – The intrigue behind the most beautiful car collection in the world (Op de Weegh & Op de Weegh)
Sleeping Beauties USA – abandoned classic cars & trucks (Marek)
To Boldly Go – twenty six vehicle designs that dared to be different (Hull)

From Veloce Publishing's other imprints:

Earthworld
Discovering Engineering that Changed the World (Edgar)

Hubble & Hattie
Dogs just wanna have fun! (Murphy)
Dogs on wheels – Travelling with your canine companion (Mort)
Emergency first aid for dogs – at home and away (Bucksch)
Exercising your puppy: a gentle & natural approach – Gentle Dog care (Robertson, Pope)
For the love of Scout – Promises to a small dog (Ison)
Life skills for puppies – Laying the foundation for a loving, lasting relationship (Millsm, Zulch, Baumber)
Partners – Everyday working dogs being heroes every day (Walton)
The Rex Factor – my dog, my friend (Gordon)
When Man Meets Dog (Blazina)

www.veloce.co.uk

First published in August 2019 by Veloce Publishing Limited, Veloce House, Parkway Farm Business Park, Middle Farm Way, Poundbury, Dorchester DT1 3AR, England. Tel +44 (0)1305 260068 / Fax 01305 250479 / e-mail info@veloce.co.uk / web www.veloce.co.uk or www.velocebooks.com. ISBN: 978-1-787114-50-0; UPC: 6-36847-01450-6.
© 2019 Chris Theodore and Veloce Publishing. All rights reserved. With the exception of quoting brief passages for the purpose of review, no part of this publication may be recorded, reproduced or transmitted by any means, including photocopying, without the written permission of Veloce Publishing Ltd. Throughout this book logos, model names and designations, etc, have been used for the purposes of identification, illustration and decoration. Such names are the property of the trademark holder as this is not an official publication. Readers with ideas for automotive books, or books on other transport or related hobby subjects, are invited to write to the editorial director of Veloce Publishing at the above address. British Library Cataloguing in Publication Data – A catalogue record for this book is available from the British Library. Typesetting, design and page make-up all by Veloce Publishing Ltd on Apple Mac. Printed in India by Parksons Graphics.

The last SHELBY COBRA

My times with Carroll Shelby

VELOCE PUBLISHING
THE PUBLISHER OF FINE AUTOMOTIVE BOOKS

CONTENTS

Foreword 5

Prologue 6

Acknowledgements 8

Chapter 1 From Boyhood Hero to Automotive Legend 10

Chapter 2 The Legend and a New Snake 16

Chapter 3 The Courtship of Carroll and Ford 27

Chapter 4 Project Petunia 31

Chapter 5 Codename: Daisy 43

Chapter 6 Shelby GR1 60

Chapter 7 Project Condor 69

Chapter 8 Shelby GTs 76

Chapter 9 Super Snakes 82

Chapter 10 Interesting Interests 91

Chapter 11 Unfinished Business 105

Chapter 12 Kismet: Daisy Comes Home 118

Appendices 138

Index 159

FOREWORD
By Aaron Shelby

Much has been written about Carroll Shelby's (Grandpa's) adventures during the Shelby American heyday in the 1960s. But, as most enthusiasts who follow Shelby recognise, those events were still early in his life, and there were many adventures and stories left to tell. Chris Theodore was fortunate to straddle a couple of Shelby's reincarnations over the last three decades. In the late 1980s, Shelby's time at Chrysler was winding down, but there was one final project that he would be involved with before departing. Theodore was the Director of Jeep/Truck Powertrain, and his team was working on a new V-10 truck engine when he was asked to show the plans to Shelby and discuss a new halo car concept, the Dodge Viper. I can still vividly remember getting to ride shotgun in the Viper with Grandpa for a lap around the Brickyard when he was the pace car driver for the 1991 Indy 500. It was the ride of my life!

In the early 2000s, having made the switch to Ford Motor Company a few years earlier, Theodore became one of the instrumental figures in bringing Shelby back into the Ford family. For me, this is one of the most exciting chapters in Grandpa's life. Notable projects included the Ford GT, the Shelby Cobra prototype (Daisy), the Shelby GR-1, and ultimately the new line of Shelby Mustangs produced by Ford. While these projects harkened back to the Shelby glory days, each represented a look to the future of Shelby's new relationship with Ford. Theodore does an excellent job of walking through the inner workings of the teams completing these projects. There was such excitement when Grandpa officially rejoined Ford. It seemed he, or a car with his name, was on the cover of one of the enthusiast magazines almost every month!

One of my great joys since joining the Board of Shelby has been hearing the stories and learning the background of Grandpa's various relationships and projects. I remember quizzing him about certain cars when we talked, but he didn't go into great detail in his conversations with me. This book brings more insight to an exciting time in Carroll's life and an important time in the life of the Shelby companies.

Aaron Shelby

PROLOGUE

Carroll's passing had left me in a deep funk. My mind was flooded for weeks with memories of our conversations. I could still hear him, and see him in my mind's eye. I slept little. When I did sleep, Carroll and I would discuss unfinished business in my vivid dreams. I decided to start writing down all the moments and conversations we'd had, so that they wouldn't slip away. Slowly the grief subsided, replaced by a wry smile as I reflected on the amazing character I had the privilege of knowing. I have had the good fortune of having many great friends and mentors over the years, from François Castaing, Bob Lutz, and Brock Yates, to lesser-known individuals like Trant Jarman, Bob Hennessey, and Vince Versage. I was fascinated by the generation that preceded me, as there was so much to learn from their knowledge and experience, but few have had the impact of Carroll Shelby.

The legendary life of Carroll Shelby is well known and has been chronicled by more talented authors. To mix metaphors, the old snake charmer was a cat with nine lives. He had cheated death so many times before that I fervently believed he would do it one more time. Carroll had done more in his lifetime than mortal men could expect to accomplish in ten. Often, his interests were so diverse that it seemed as though he was living two or three lives simultaneously.

So what can I contribute to this story? Over nearly a quarter century, Carroll and I had become very close. Between visits we would talk at least every other week, usually twice a week. I came to the realization that there had never been a first-hand account of what it was like to be with Carroll in his later years. Moreover, few people knew what really went on behind the scenes regarding Carroll's involvement with Chrysler and Ford, other than his appearances at corporate press events. Then there were all the projects, hopes, and dreams we shared together.

Why was I so fortunate to be included among Carroll's amazing circle of friends? The last years of Carroll's life were filled with energy and accomplishment. He wasn't one to look backwards. Yes, he enjoyed getting together with his contemporaries, entertaining the crowd with his fabulous storytelling, but his curiosity and drive kept Shelby focused on the future. It finally occurred to me that we became close because our discussions always centered on future products, technology, and dreams. Carroll was never done, and I wanted to be part of that journey into the future. This focus on the future certainly contributed to Carroll's longevity.

A few months after starting this book, I suddenly stopped, due to some unsettling circumstances. I had given a two-hour interview to a journalist concerning a special period in my career, and did my best to give credit to all the people who had created such a wonderful work environment. I thought nothing of it, until I received word that I had upset some of my colleagues. I had never seen the finished article, and went back to see how I had offended. Yes, there were a couple of minor technical errors, but what I suspect upset others was the editor's introductory paragraphs in the multi-part article that, indeed, made it seem like I was taking credit for all that had been accomplished. Horrified, I apologized to those who were offended, but also learned that vivid memories are not always correct. People I know and trust remembered a different outcome from the same event! It reminded me of the old Indian parable of the four blind men and the elephant, each describing the elephant from their own perspective, distrusting the description of the other.

What follows is a first-hand account of my time with Carroll. Where my memory was fuzzy, I made sure to get corroboration from others. I thank all those who reviewed the chapters, helped me check my facts, and filled in missing pieces of the plot. Several times, I learned new things regarding the back story to an event I witnessed, and I have included interesting details from people that were there. This

PROLOGUE

book chronicles the truth about my time with Carroll, as I know it.

I am not an author or journalist; my biggest fear is that in telling this story, I fail to capture the essence of this colorful, amazing man – his humor, laugh, storytelling, drive, curiosity, intelligence, and humanity. Carroll was no angel, but his flaws only made him that much more memorable. I hope that Carroll's own words from our many conversations will reveal his true character, and that you will enjoy getting a little closer to the real legend.

ACKNOWLEDGEMENTS

First, I must thank Edsel Ford II. He is the quiet hero, without whom the Shelby legacy would be incomplete. Not only was Edsel instrumental in bringing Carroll back to Ford, but also helped ensure the return of Shelby badged Mustangs and secure the future of Shelby high-performance "sport" cars. I thank Edsel for the time he spent with me reliving history, and his kind note of encouragement regarding this book.

Team Shelby was greatly instrumental in providing additional detail. Carroll's grandson, Aaron Shelby, read and suggested changes, provided insight into his grandfather's character, and I thank him for the thoughtful forward. Gary Patterson, President of Shelby American, reviewed the manuscript, and provided additional details to make the story accurate and complete.

John Clinard, Ford's extraordinary PR spokesperson, was incredibly helpful in helping me track down the true story about Carroll's return to Ford. One of the most knowledgeable historians in the business, John helped make sure all the dates and names were correct, and provided fresh leads, in addition to his immense help in editing. Jim Farley provided the details of securing the Shelby/Ford legacy, reviewed the relevant chapters and related his "favorite story" about Carroll and Ulrich Bez.

My good friend, Bill Warner, reviewed every page multiple times, added stories, and provided rare photos of Carroll's early years. His gracious invitation to bring Daisy to the Amelia Island Concours d' Elegance provided the impetus to get Daisy running once again.

Long-time colleague, *Motor Trend* technical editor, Frank Markus, taught me the 'rules' of automotive journalism. Asked to take a quick look at the draft manuscript, Frank instead provided a detailed edit.

Regarding Carroll's years with Chrysler, Roy Sjoberg, Dick Winkles, John Fernandez, Charlie Henry, and Neil Hannemann filled in the missing details of Shelby's involvement, and related entertaining Viper stories and photos. Bob Lutz confirmed the many stories attributed to him over the years, adding fresh insight. Finally, a special thanks to Joe Gall for collecting the names and responsibilities of the original Viper team members.

The history of the courtship of Carroll and Ford was finally documented with the help of William Jeanes, Keith Crain, John Coletti, and John Clinard. John Coletti went on to fill in the blanks regarding the Ford GT and Shelby GT500 programs. The success of the Ford GT would not have happened without the leadership of Fred Goodnow, his assistance (including the Ford GT Team history yearbook) was invaluable. Special thanks to Neil Ressler for providing additional background on Petunia, and to Larry Edsall for providing photos from his book.

The details of Carroll's involvement in Daisy might have been lost to history without Bud Brutsman's documentary, *Rides*. Thankfully, Bud took over a thousand still photos during the vehicle's development, and Cynthia Whorton, director of operations at BCII, spent untold hours digitizing all of them for posterity. They brought back memories of the wonderful times we had with Carroll. In addition, the Advanced Product Creation team, under the direction of Hermann Salenbauch and Manfred Rumpel, documented the progress of Daisy and the GR1 with detailed build books. Thanks go out to Bob and Mike Nowakowski, Gene Duprey, and Brian Hermann for getting Daisy running once again, and the rest of the Technosports team involved in fabricating and assembling Super Snake II.

Without Harley Cluxton III, there might have never been a Super Snake II. His restoration and sale of the original Super Snake provided the inspiration and encouragement for undertaking the project with Carroll. A special thanks goes out to the wonderful Evi Gurney for her insight, and for approving the use of Dan's speech at the Shelby Tribute.

ACKNOWLEDGEMENTS

Jason Vines encouraged me to write this book, and set an example with his own writings, including *What Did Jesus Drive* and *The Last American CEO*. I treasure his advice and counsel.

'GT Joey' Limongelli reviewed every page and replied "More pictures!" Joey was right, the Shelby story would be incomplete without the many photographs and images provided by Bill Warner, Larry Edsall, John Coletti, Bud Brutsman, Aaron Shelby, Scott Black, Howard Walker, Tomy Hamon, Don Wood, David Friedman, Claude Haycraft, John Fernandez, Dick Winkles, Ken Gutowski, Dave Byron, Mike Luzaders, and Mike Zevalkink that augmented my own files. Shelby American, Ed Loh and the *Motor Trend* team, Mike Fazioli at *Road & Track*, Ford Motor Company and Fiat Chrysler Automobiles (FCA) all contributed to the photographic history.

This book would not have happened without the support of Tim Nevinson and the entire Veloce publishing team, especially Kit Anderson for their thoughtful editing. As a first time author, I learned to appreciate the creativity and immense effort required to publish a finished work.

Finally, the unsung heroes are all the designers, engineers, technicians and fabricators who made these fabulous vehicles a reality. I have tried to capture in the Appendices as many of the names of the various team members as possible. My apologies to those I have missed, but you know who you are. You should be proud!

Chapter 1
FROM BOYHOOD HERO TO AUTOMOTIVE LEGEND

We all have heroes when we are growing up. Usually they tend to be sports, music, TV, or movie stars. I suspect I was a bit different than most. I loved cars and anything associated with them. My dad caught me jacking up his car to change the tire when I was four or five years old. My older cousin provided babysitting services by driving around parking lots, and having me name the year, make, and model of every car I saw. Seeing my first soapbox derby instilled a love of racing. Back then, car dealers kept new model introductions secret by papering showroom windows, so my dad would take me to the factories in Detroit in order to get a sneak peek through the storage lot fences. I collected car brochures and restyled the new models, read *Rod & Custom* and customized car models.

Perhaps a bit atypical for a kid from Detroit, I was more interested in sports cars and road racing, than muscle cars on Woodward Avenue, or dragsters down at Milan Raceway. Consequently, I read *Sports Car Graphic*, *Road & Track* and everything written by Ken Purdy and others, glorifying great drivers, present and past.

Rather than baseball or football stars, my heroes were the great race drivers of the times. Every Memorial Day I'd listen to Chris Economaki's radio broadcast of the Indy 500 and root for Rodger Ward, AJ Foyt, Parnelli Jones, and later, Jim Clark. The only American champion in Formula 1 at the time was Phil Hill, competing with the likes of Jack Brabham, Jim Clark, and Graham Hill – all chronicled by Rob Walker in *Road & Track*, plus the legends of Fangio, Nuvolari, and Caracciola that had been glorified in the racing books I devoured.

In road racing, I no doubt read about Carroll Shelby's 1959 win at Le Mans with co-driver Roy

Shelby at Le Mans in 1959. (Courtesy David Friedman)

Shelby and David Brown celebrate victory at Le Mans. (Courtesy *Motor Trend*)

Shelby and Moss studying the Sebring course. (Photo by Claude Haycraft. Courtesy Bill Warner)

Shelby wins the LA Examiner GP at Riverside, April 1960.

Shelby: driver of the year. (Courtesy *Motor Trend*)

Shelby and Gurney at Riverside winners' dinner, 1957. (Both Courtesy David Friedman)

Shelby's last race at Laguna Seca, October 1960. (Courtesy David Friedman)

THE LAST SHELBY COBRA

Salvadori, but probably only learned about Shelby's other racing accomplishments after the fact:
- 1956 and 1957 SCCA National Champion.
- 1956 Sports Illustrated Driver of the Year and New York Times Driver of the Year.
- 1960 USAC road Racing Champion.
- Finishing 2nd overall in his last race at Laguna Seca, while popping nitroglycerine pills for angina.

Of course, I had many road racing heroes, including the likes of Stirling Moss, Phil Hill, Pedro Rodriguez, and later, Jim Hall, Bruce McLaren, Dan Gurney, and Mark Donohue. For a budding 'car guy' it takes much more than being a racing great to become a legend.

Car guys can remember their first sighting of a car that made them melt – be it a photograph on a magazine cover, or in the flesh. Cars like the 1961 Jaguar E-type, 1962 Shelby Cobra, 1962 Ferrari GTO, or 1963 Corvette Stingray. Those were the cars we lusted after and dreamed of owning, driving, and perhaps even racing someday. To bring those dreams a little closer to reality, there was the 1965 Shelby Mustang GT350, and later the GT500 models. Performance cars that you might someday be able to afford.

Car guys would argue which make was the best, but the racetrack, of course, was the ultimate discriminator. Carroll Shelby's Cobra started by first

FROM BOYHOOD HERO TO AUTOMOTIVE LEGEND

knocking off the American champion, Corvette, in US Road Racing by winning three championships and its first international FIA race at Bridgehampton in 1963. Carroll's Daytona nearly won the FIA World Championship in 1964, were it not for Enzo Ferrari's intervention with the FIA rules committee to take the Monza race out of the championship schedule. This was also the year the Daytona finished first in the GT class, and 4th overall at Le Mans. In 1965, Shelby won the FIA World Championship with the Cobras and Daytonas, but the Team Shelby Ford GTs failed to win at Le Mans, setting the stage for 1966 ...

The seminal event for youthful road racing enthusiasts like myself came in 1966, when the Ford GTs finished 1-2-3 at Le Mans, finally beating Ferrari on its home turf (and won again the following three years!). This accomplishment meant that Carroll Shelby was the only person to have ever won Le Mans as a driver, manufacturer, and team owner, a feat never repeated. For American sports car enthusiasts brought up on the notion that only Europeans could build proper sports cars, this win reaffirmed our national pride. America could compete with the best in the world and win! Three years later Neil Armstrong set foot on the moon, indelibly etching that notion in the minds of all Americans.

Despite no US broadcast of the Le Mans race, news of Ford's victory quickly spread. Car crazy teens like myself switched to GT40, Shelby GT350, Cobra, and Daytona slot cars, as we imagined vanquishing the all-conquering Ferraris. Carroll Shelby was now an

Opposite Finishing the first Cobra in Dean Moon's shop, February 1962. (Courtesy David Friedman)

Jacque Passino, Don Frey and Shelby receive the 1966 FIA Constructor's Award. (Courtesy David Friedman)

Shelby with the 1967 Le Mans winning Ford GT Mk IV. (Courtesy David Friedman)

automotive legend. He had conquered the automotive world by the age of 43! Perhaps Jay Leno put it best, upon seeing the May 1965 cover of *Road & Track*, "I want to be that guy!"

What makes an automotive legend? Upon a little reflection, the first three criteria are relatively straightforward: 1. Great race car driver; 2. Design, build, and sell your own car with your name on it; and 3. Race it and beat the best in the world! When the industry was in its infancy, this was the formula used by Louis Chevrolet, Henry Ford, and a host of Europeans to start and grow their car companies. In the post WWII era, I can only think of two who have succeeded, Enzo Ferrari and Carroll Shelby. Many others, from Preston Tucker to John DeLorean have tried and failed. Of course, there are many other great marques, including Porsche, Aston Martin, Lotus, and Lamborghini, but they do not have the provenance to meet the three criteria. Other great race drivers have met at least two of the criteria, and are indeed racing legends. Perhaps the best being the late Dan Gurney, who was not only an outstanding driver and human being (and largely responsible for the 1964 Daytona and 1967 Ford GT Mk IV Le Mans wins), but the only American to win an F1 race in a car of his own design and manufacture. No wonder he was *Car and Driver*'s candidate for US President! Jim Hall went on to have great success with his Chaparrals and race teams. Roger Penske became an industry mogul and probably the winningest team owner of all time, while Bobby Rahal, Mario Andretti, A J Foyt, Jackie Stewart, and others went on to manage successful race teams.

FROM BOYHOOD HERO TO AUTOMOTIVE LEGEND

All car guys, myself included, secretly dream of designing and building their own cars. Moreover, they dream that their car will be the best in the world at what it does. Carroll Shelby was one of the few to realize such a dream, and became an American hero among car guys.

I believe, however, that there are two more criteria to cement one's legacy as an automotive legend. The fourth is a larger than life personality. Enzo Ferrari might best be described as being imperious and Machiavellian. Why else would was he called 'Il Commendatore'? Enzo had only been a mediocre race driver, but possessed the managerial and political skills to create a conquering Scuderia, and ultimately the bespoke brand of automobiles that carried his name. Carroll Shelby, on the other hand was a risk taker, engaging and funny. His East Texas drawl, 'aw shucks' down-home self-deprecation, just-do-it attitude, and public persona was something that Everyman could love. He was equally at home with barons, movie starlets, business executives, race drivers, mechanics, and the man on the street. He had a magical way of engaging and connecting with them all. Carroll wasn't an engineer or car designer. Nor was he a detail-oriented manager. He had the knack to seize an opportunity, attract and select the right people, and turn them loose to accomplish seemingly outlandish goals.

The last criteria to becoming an automotive legend is longevity. Enzo Ferrari lived long enough to see his cars dominate the racing scene and his company become the premier sports GT manufacturer.

Carroll Shelby had accomplished both racing domination and built a legendary sports car by the time he was in his early 40s. He stopped building Cobras and Shelby Mustangs by 1967. Carroll once told me, "I never made any money on the Cobra," (although I suspect more accurately, Carroll plowed profits back into his racing efforts). Carroll quit the car business and spent the next decade in Africa as a bush pilot, guide, and big game hunter. With a failing heart, Carroll nearly died young, and perhaps the legend might have waned. Upon his return from Africa, however, his old pal Lee Iacocca lured Carroll back into the business to inject some excitement into Chrysler products. The legend took on new life for the next generation of enthusiasts.

Shelby's accomplishments have been chronicled by authors far more capable than I, and in far more detail than the preceding summary of his first 65 years. The following chapters are about my personal relationship with Carroll, one that grew into a deep friendship over the years. They provide an insider's account of Carroll's return to prominence in the industry with Chrysler and Ford, his continued focus on the future, his unique personality, and the many unpublicized projects he continued to work on, up until his passing.

May 1965 cover photo from *Road & Track*. (Courtesy *Road & Track*)

Chapter 2
THE LEGEND AND A NEW SNAKE

In 1978, while working on the Chrysler/Calspan Research Safety Vehicle at Creative Industries, word came down that Lee Iacocca (fired by Henry Ford II) had joined Chrysler. I was not surprised. For days, Creative personnel had been polishing and primping a navy blue Chrysler Fifth Avenue for an unnamed dignitary. One of my colleagues got the job to pick Lee up at his home, and deliver him to Highland Park. The rest is history. Lee went on to save Chrysler, paid off government-backed loans, and became corporate pitchman ("If you can find a better car ..."), national hero and best-selling author.

Had I stuck around, I might have met Carroll Shelby six years earlier, but in 1980, I made the fateful decision to help start up a turbocharging company, Legend Industries, in Hauppauge, NY. The only problem was that our two clients were Fiat and DeLorean. Within two years, Fiat withdrew from the US market, and John DeLorean got busted on a cocaine deal. Without customers, we went bankrupt!

At the same time, some of Chrysler's product successes with the K-cars had started to fade. Lee Iacocca sent word to Shelby that he wanted to use some of Carroll's magic to spice up the product line. With nothing but four-cylinder, front-wheel drive small cars in the line-up, Carroll was reluctant, but felt obliged to his old friend. In 1982, Chrysler announced the hiring of Carroll Shelby as a 'performance consultant.' The Shelby name was still big. Charlie Henry, who was Marketing Manager of Chrysler performance parts recalls the 1982 SEMA show: "As I was talking to Carroll in the booth, my wife arrived with the baby in a stroller. I said, 'Carroll, this is my wife, Linda, and my daughter Shelby.' He lit up like a Christmas tree and said, 'Well, Ah'll be. Ah'm gonna call mah drinking buddy, Edsel Ford, and tell him, you looky here, Ah got people namin' babies after me and nobody is namin' their kid Edsel anymore.'" He went on to ask for a picture of the baby to hang on his wall. In fact, Shelby plastered his office with pictures of kids named Shelby or Carroll!

While I wasn't at Chrysler at the time, many young Chrysler engineers became involved with the volunteer Team Shelby race effort: John Fernandez, Neil Hannemann, Dick Winkles, Pete Gladysz, Ken Nowak, Don Jankowski, Ray Schilling, and Al Fields, among others. All would play a part in the ongoing Shelby story. Fernandez went to work at the Chrysler Shelby Performance Center in Santa Fe Springs, California. Within a year, Carroll founded Shelby Automobiles Inc, with its first products being the 1983½ Dodge Shelby Charger, and 1984 Dodge Omni GLH (Goes Like Hell). A small volume assembly plant was set up in Whittier California to produce Shelby badged models (as distinct from Dodge models), the first of which was the Shelby GLH-S (Goes Like Hell-Some more). To introduce the GLH-S, Shelby scheduled a daylong media event at Willow Springs Raceway with two vehicles, one with a roll cage. The first of eight journalists scheduled to drive, a bit overconfident in his abilities, lost control in turn eight and flipped the car numerous times as the Shelby crew watched in horror. Carroll walked over to Fernandez and said, "Better go out there and see if they're OK" and after a moment's thought added, "and if they are, kill'em for me!" After that inauspicious start, the team retrofitted the second car with the roll cage from the wrecked GLH-S and the drives continued, generating great press. Other Shelby models followed, based upon the Dodge Omni, Charger, Lancer and Shadow, even a Dakota pickup fitted with a V-8. Production of these limited edition cars ended in 1989, although Chrysler continued to sell Dodge-branded Shelby models until 1991. The Team Shelby racing effort won multiple endurance championships, with Neil Hannemann, Charlie Henry, and others sharing driving duties.

With Shelby production nearing an end, Fernandez joined Chrysler to head Vehicle Development on the Neon program. To assist in the transition, John helped secure Chrysler's new 3.3L V-6 engines for Carroll's SCCA Can-Am spec car. John had also been working

THE LEGEND AND A NEW SNAKE

Carroll with a Shelby GLH-S. (Courtesy *Motor Trend*)

with Carroll on a prototype 2900lb V-8 sports car. The preliminary design was complete, and fabrication of a prototype started, but the project aborted when Carroll became an adviser on the Viper project. The rest of the Shelby operation shifted gears to providing 'performance research' services for Chrysler. Neil Hannemann, a bright young engineer, and talented endurance racer, led the effort, working on everything from advanced turbocharging to AWD. Dick Winkles became the high-performance engine guru, developing production Turbo II, III and IV engines as well as supporting racers with higher performance derivatives.

I went to work for American Motors in 1985, first as Chief Engineer of Advanced Vehicles, and later as Director of Engine Engineering. A mutual friend introduced me to François Castaing, the brilliant and charming French VP of Engineering who had been instrumental in Renault's entry into Formula 1 with the, then radical, turbocharged engines. We hit it off immediately, and I found myself working with a small but dedicated team of engineers on such products as the Jeep Grand Cherokee and new 4.0L fuel-injected I-6 engine. American Motors housed its engineering operations in a classic Albert Kahn designed building in Detroit that had been home to Nash and Kelvinator. Central to the beautiful façade was a four-story structure that enclosed a water tower. The building had an ambiance of times gone by, where the likes of George Mason and George Romney had once ruled from the second floor corner office. Hidden in back was a huge engineering facility with prototype shops, test equipment, garages and dynamometers, modernized during the era of Renault's ownership of American Motors.

Although AMC had seen tough times, things were on the upswing as Roy Lunn, François Castaing, Jim Julow and others created the Jeep Cherokee and established the booming four-door SUV market. Across town in Highland Park, Lee Iacocca had rescued Chrysler and profits were rolling in. Iacocca went on a buying spree with acquisitions of Gulfstream, Lamborghini and others. Lee wanted to add the iconic Jeep brand to Chrysler's expanding portfolio and began secretly negotiating with Renault to purchase AMC. On March 9, 1987, Chrysler made the surprise announcement that it was acquiring AMC, and I soon found myself working for my former employer. Rather than return to Renault, François decided to cast his lot with Chrysler. He had

John Fernandez (left) observes a driver change on a Shelby Can-Am car. (Courtesy John Fernandez)

quietly formed an alliance with Bob Lutz, who had also recently joined Chrysler from Ford, to head Chrysler's truck group. As the Chrysler and AMC teams merged, Lutz rose to become President of the Company, with Castaing taking over Engineering as Executive Vice President.

Shortly after the AMC deal closed, the stock market crashed on Black Friday, October 17, sending Chrysler into another one of its near death experiences. In a declining market, with an obsolete line of K-car products that had been sliced and diced into every conceivable configuration, Chrysler was in trouble. Iacocca's taste for baroque styling with padded vinyl tops had clearly gone out of favor. To his credit, Iacocca entrusted 'product guru' Lutz with the task of saving Chrysler once again, with a new product onslaught. Faced with impending disaster, Lutz managed to pull the team together to save the company. It was an exciting time, which I later came to call the 'Camelot' years.

An aggressive product plan was proposed, beginning with a new Grand Cherokee in 1991, LH large cars in 1992 (deridingly called "last hope" by some pundits), new Ram pickup, followed by PL Neon, NS minivans and JA mid-size Cirrus, Stratus and Breeze cars. With sales falling, the company was hemorrhaging red ink and facing an increasingly critical press. Lutz knew he had to do something to demonstrate that a brighter future was on the horizon. Reportedly, one day, while taking his Autokraft Mark IV replica Cobra for a ride along the winding roads surrounding his Saline, Michigan home, Lutz came across the solution: if Chrysler could quickly introduce a halo sports car, the public perception of Chrysler would improve, increasing anticipation and interest in the other new products that were to follow.

I first met Carroll in the spring of 1988 while serving as Director of Jeep/Truck Powertrain. We were developing a new line of fuel-injected Magnum V-6 and V-8 engines to replace the anemic carbureted B-block motors that Chrysler had relied upon for years. Also on the drawing board was an 8.0L V-10 engine to power a new line of Ram pick-ups. Early one morning, François called and said, "I need you to come down to the boardroom after lunch and meet with Carroll Shelby. Lutz and I have been discussing doing a new sports car, and I want you to show him the new V-10 engine." With the exciting news of the opportunity to meet the legend himself, and work on a new sports car, I collected up all the V-10 drawings and specifications, and waited for the appointed meeting.

The boardroom was around the corner from François' office. It was a typical 1930s rectangular conference room with mahogany paneled wainscoting and a huge mahogany table surrounded by old brown leather armchairs. Shelby and François were already there when I arrived. John Kent, who was responsible for the upcoming Wrangler, joined us. Carroll was taller than I expected, and a little heftier than the memories of the lanky legendary racer etched in my brain. Now 65 years old, the wavy black ducktail was replaced by a more businesslike cut, which was more salt than pepper, and definitely silver along the temples. I recall a dark cardigan sweater, and large black Oxfords with thick rubber soles.

"Hello Chris" he said with his homey, coarse East Texas drawl, as if we'd known each other for years.

"It's an honor to meet my boyhood hero," I replied.

François excused himself for another meeting, and Carroll picked up the conversation. "Whatcha got under your arm?" inquired Carroll.

"Drawings of an engine that just might suit your needs," I replied.

I unrolled the drawings on the long conference room table while Carroll proceeded to outline the plan. "Lutz and I have been talking about doing a modern version of the 427 Cobra. John Fernandez and I have been fooling around with a sport car of our own, but I don't have the resources to do it, so we decided to join forces." I noted that Carroll said "sport" car, rather than the conventionally recognized "sports" car, but, when you think about it, he was probably right – they are only used for one sport.

"Well, the old Cobra used a 7.0L V-8. How about topping that with an 8.0L V-10?" I asked.

Shelby seemed a little concerned about size and weight, but I showed him the drawings and suggested we could develop an aluminum block version, which would weigh less than the old 427 Ford. "We should be able to get 400hp and at least 400lb/ft of torque. That would make it the most powerful car in production." Carroll bought into the proposal for an aluminum block, and as the discussion continued, it was clear that we were on our way. John Kent provided drawings of the Jeep YJ frame, on the off chance that it could be useful.

I left the meeting on cloud nine. We were going to do a car with Carroll Shelby, and make history again!

I didn't hear much more about the project until the fall. The design office asked for drawings of the V-10, so they could make sure it fit in the concept car.

Late that fall, Al Turner, who managed special projects for the design office, came by asking if a V-10 would be available. While prototype parts were coming in, the first engine would not be ready in time. Consequently, Al had Roush Engineering furnace braze together two 5.9L engine blocks to create a running V-10 for the show car.

The PR department scheduled a press preview at the Highland Park Design Dome before the opening of the 1989 NAIAS. Giving the press a scoop before the opening of the show was the only way a 'beleaguered' Chrysler could break through the media clutter, and Viper was the lightning rod to show that Chrysler was on its way back with a new line of exciting products. I hadn't seen the sketches or the clay, but when I saw the show car as it rolled off the truck and into the staging area, it was love at first sight. We had to do this car, and a skeptical press thought so, too.

Viper was the hit of the auto show, and helped make the Detroit show one of the highlights of the international circuit. Customer response and press coverage was phenomenal. Viper was on the cover of every major magazine, including an exclusive review with Jack Keebler, Detroit editor of Road & Track. Letters and checks were coming in. As the 'four fathers' Lutz, Shelby, Castaing and Gale posed for pictures, they chuckled that all they had to do was figure out how to build it. In the following months, they hatched a plan to get the production Viper program approved. François would form a team of enthusiast volunteers that would act as a 'pilot' for the platform team concept (even though several of the teams already existed). Engineers would be 'lent' to the Viper program, others volunteered with the understanding there were no guarantees their old job would be waiting upon completion of the project. Tom Gale, Executive Vice President of Design, would start working on a production-feasible clay and Tom Stallkamp, EVP of Purchasing, would ask Chrysler's key suppliers to support the program by absorbing tooling costs. Shelby's role, as he put it, was to "tell Iacocca that doing the Viper was a good idea, and that it wouldn't cost Chrysler a thing, because the suppliers were going to pay for it."

The task of getting an aluminum V-10 completed in time for Viper production took some interesting turns. The new truck V-10 was the brainchild of Lutz: "Just graft a couple of cylinders onto the 360 V-8." In truth, there was little technical reason for doing a V-10. While it just requires two more cylinders, the fact of the matter is that the expensive tooling has to be all new (block, heads, etc) so investment savings were minimal. Starting with the 360, however, did reduce development costs. Yes, V-10s were the engine of choice for Formula 1 at the time, but that would be of little notice to good-old-boy truckers. Still I was a strong supporter of the V-10, for the simple reason that it was different and would attract inordinate interest. The new Ram pick-up was going to attract a lot of attention. Rather than doing the 'me-too' pick-up that Chrysler had originally planned, the new Ram would take on the macho appearance of a Mack truck – one that would scare the daylights out of anyone checking their rear view mirror as it bore down on them. A V-10 as the top-of-the-line engine would give truckers bragging rights: "Mine is bigger than yours."

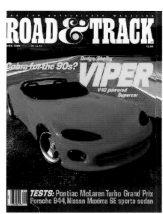

April 1989 Viper cover story. (Courtesy Road & Track)

My job was to get the V-10 engine approved. With a broken leg in a full-length cast, Lutz couldn't maneuver into the cramped program approval room. As I approached the room to pitch the V-10, Lutz, stretched out on the bed of an electric cart parked in the hallway, pulled me aside and said, "Don't come out of that room without an approval!"

Fortunately, after the V-10 pitch, there was only one question. "Will it fit in a LeBaron?" asked Joe Campana, the 400lb+ VP of Marketing.

"Only if you want it to stand on its nose," I replied. Everyone laughed as Campana's face reddened, and I escaped with my job intact.

I had hoped that our engineering team would do the aluminum version of the V-10, but a fortuitous set of circumstances changed those plans. Chrysler had purchased Lamborghini in 1987, and decided to participate in Formula 1 to burnish its brand image. Mauro Forghieri, the genius behind all of Ferrari's race cars from 1962 on, joined Lamborghini to design an F1 V-12 engine, as part of the new Lamborghini Engineering Division. Mauro was responsible for creating the legendary Ferrari GTO, inventing the first airfoil rollbar, and creating the winning Formula 1 series of 312 racers (including the invention of the transverse gearbox in the 312T). Legend had it that Mauro installed a drafting board next to his mistress's bed, so that when he woke up with an idea, he could put it to paper. Reportedly, the transverse gearbox resulted from one of his dreams.

In the F1 days, François and Mauro were collegial competitors. Now, François asked the Powertrain Tech Club to visit Mauro at the new facility in Modena to learn F1 technology from one of the masters. Mauro was engaging and the V-12 engines and prototype shop impressive. The engine was powerful, but Mauro needed assistance with the electronics and calibration. The Tech Club assigned Dick Winkles and Lee Carducci to help get the most out of the engine. In less than a year, the Lamborghini F1 engine became the second most powerful in racing – quite an accomplishment for such a small operation!

Although Lamborghini had made great progress in F1, the program ended when Ron Dennis, team principal of McLaren, backed out of a deal to use the motor in their 1994 F1 car despite its having shown great potential. The unfortunate saga is too complicated to detail here, but proved beneficial to the Viper program.

Lamborghini Engineering, led by Forghieri, now became an engineering service. They secured a contract for an electric vehicle program, but François thought it might also be a good idea to try Mauro out on the aluminum Viper V-10. We sent the drawings over to Forghieri, and asked him to come back with a proposal. Mauro's design impressed us with its incorporation of F1 engine technology in a classic American pushrod engine design. He proposed very short water jackets, unique external port cooling system, and engine block with bedplate support for the crankshaft. Lamborghini Engineering received the contract to finish the design and provide the first few prototypes for the Viper program, as they could provide parts far more quickly than Chrysler's domestic suppliers. I doubt Carroll Shelby knew that his old nemesis from the 'Cobra/Ferrari Wars' had been involved in the Viper program, and it was probably just as well that he didn't!

The plan of the four fathers worked, with the formation of a Viper Feasibility Team, approved on March 28, 1989. François selected Roy Sjoberg, to lead the team. Roy had joined Chrysler in 1985, after a 25-year career at General Motors. His experience with composite materials, skunkworks projects, and inclusive management style was just what the cross-functional team would need. Roy solicited volunteers from within the ranks of Chrysler 'car guys,' including most, if not all, of 'Team Shelby.' The feasibility study convinced management that Viper could be profitable, while Lutz told the board that all the free publicity justified the $50M investment. The board approved the program, and Chrysler announced on May 18, 1990 that the Viper would be in production within three years of the concept's introduction.

My role in Viper was minimal: mainly to provide support for the powertrain effort. Dick Winkles and Jim Royer, a seasoned, silver-haired engine design veteran, were 'loaned' to the team, along with 'Hemi NASCAR' dyno cell 13, and operator, Bob Zeimis. To tackle the job in record time, Sjoberg colocated the new team, of about 50 engineers, in some abandoned space at JTE, the old AMC facility, and they were off to the races. The team designed and fabricated a suspension mule, followed by a design-representative prototype within nine months. Due to the tight timing, a V-8 powered the first Viper prototype. Finally, in April 1990, the first V-10 Viper prototype was running. Dick and Jim soon found that some of Forghieri's F1 practices didn't translate to a road car. In particular, the aluminum crankshaft bedplate simply did not work, and was replaced with conventional steel bearing caps. The engine also ran hot with the short cooling jackets, requiring the addition of an oil cooler. Thanks to a dedicated effort, however, they redesigned the V-10 to produce prodigious power and torque with bulletproof reliability. Sadly, Jim passed away a few years after the Viper was in production.

To keep the program on track, Sjoberg organized quarterly Viper Technical Policy Committee (TPC) meetings with the four fathers and supporting executives, who could break barriers and make sure the program stayed on track. Viper was to be a brutal, Spartan sports car: it had no side windows, exterior door handles or high tech content that could detract from performance or make it more difficult to execute in such a short time frame. I attended as often as possible, and recall some of the debates. One of the most controversial was about the inclusion of optional air-conditioning for the 1993 models. Its addition, despite being contrary to the program charter, proved fortuitous, as the foot box tended to get very warm with its close proximity to the catalysts and side pipes. Lutz often told the story about the debate regarding the adequacy of the insulation around the side pipes.

THE LEGEND AND A NEW SNAKE

Viper team photo with the first V-8 powered prototype. (Courtesy FCA)

The issue was settled at the test track, when the team watched François exit the Viper with his pant leg melted and up in smoke!

Carroll's role at the TPC was to be the spiritual conscience of the car. As Sjoberg recalls, "His biggest concern was weight, and that Viper should be as light as the Cobra."

During one of the TPC meetings, Carroll was standing in the corner of the conference room as a heated debate raged around the table. I sidled up to him and whispered, "What do you think?" In one of the biggest understatements of all time, he replied, "Aw Chris, I'm just an old marketing guy. You guys are the experts." True, Carroll had long ago learned the benefits of marketing with his trademark bib overalls, and 'just a chicken farmer' humility, but he was very much more than that. Years of racing and building cars with the likes of Phil Remington and Ken Miles had taught him what it took to build a great "sport" car. As Carroll's heart deteriorated, his visits with the team became infrequent. Roy recalls that, "we had to pick him up at the parking lot of JTE and drive him back to the team center. He was really short of breath!"

In the summer of 1990, Chrysler held its long-lead press preview in Sedona, Arizona. New for the 1991 model year were the 300hp AWD Dodge Stealth and 224hp Dodge Spirit R/T powered by a turbocharged 2.2L that we had developed using Lotus designed DOHC four-valve heads. Journalists flew in to Phoenix, where they picked up a test car and drove (raced) to a luxury resort. Of course, the police pulled Lutz over for speeding. I can still see him posing next to the R/T, one elbow on the roof and the other hand holding a Cohiba cigar, as the slower journalists drove by. The ticket didn't stop Lutz from further high-speed exploits. He was later stopped in the Viper with Winkles as co-pilot. "I clocked you at 127mph," said the officer. "Hi, I'm Bob Lutz, President of Chrysler, how do you like my new sports car?" responded Lutz as he shifted into sales mode, "Want to see the engine?" After popping the hood to reveal the V-10, the officer agreed not to write Lutz up, under the proviso that he drive the Viper back to the police station for all to see.

The next day included a tech review, a drive through the countryside, and time for photography against the beautiful Sedona backdrop. After an outdoor barbeque dinner, the event was wrapped up with cocktails at the second-story bar, watching the Detroit Pistons and Chicago Bulls in the play-offs. Chrysler's PR manager, the late Tom Kowaleski, invited everyone onto the veranda for a toast. In the background, one could hear the rumble of a car approaching. It was the first V-10 prototype Viper, with Lutz at the wheel. He did a couple of laps around the courtyard as the press

Lutz and Winkles at the Mormon Lake Police Station. (Courtesy Dick Winkles)

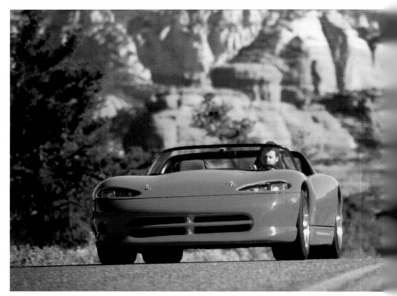

Winkles driving the Viper prototype VM02 around Sedona. (Courtesy Dick Winkles)

looked on, and then stopped so they could inspect his new toy. Point made – there was more to come from the new Chrysler! When the party wound down, Dick Winkles leaned over to me and whispered, "Want to go for a ride?"

Off we went to where Dick had hidden the Viper. He fired it up and we drove into the cool Arizona night, with the vibrating, misaligned Marty Feldman headlights pointing cross-eyed at the millions of stars in the black sky. We raced around the local roads as the exhaust pulsed from the V-10's side pipes echoed off Sedona's red rock formations. It seemed as if the writings of Ken Purdy had come to life.

After the heart transplant, Carroll was like a new man: "I got the heart of a 34-year-old." As Sjoberg described, "He burst into the team center after walking all the way back from the lobby. What a difference!"

It was a good thing that Carroll was back, as he helped turn one of Chrysler's rare PR snafus into a PR extravaganza. Dodge Marketing had arranged for the Stealth to be the pace car of the Indy 500. There was an uproar when the UAW and American loyalists found out; no way could a Japanese-built car pace America's most important race! To defuse the issue, Lutz directed the Viper team to move up the build of a production prototype to pace the Indy 500. Pulling a rabbit out of their hat, the team completed the prototype and a backup just in time for the Memorial Day classic.

Carroll was to be the pace car driver. He relished the job, giving rides around the brickyard to anyone and everyone at Indy's 'Month of May.' Each morning, he would cook up a different batch of chili for the support crew to enjoy at lunchtime or throughout the day. At one point, the PR people reported the Viper stolen, only to find out that Shelby was out on the track again. Jim Grady, who headed sales at Borg-Warner, got the scare of his life as Shelby circled the track in the rain, one hand on the wheel, the other wiping down the inside of the windshield with a towel. Johnny Rutherford had instructed Carroll to hug the wall as he went around, and Jim thought he was within inches of disaster. Better yet was the ride for General Norman Schwartzkopf. After the ride, he asked, "Who was that guy?" When told that Carroll was a recent heart transplant recipient, Schwartzkopf commented that he had been in many wars, but had never been as scared as that ride around Indy. The best, however, was when Shelby took C Van Tune of *Motor Trend* for

Shelby and the Viper Indianapolis 500 pace car. (Courtesy Shelby American)

Carroll with his grandson, Aaron, at Indy. (Courtesy Aaron Shelby)

a ride. While circling the track, Carroll groaned and clutched his chest. As Van looked on in horror, Carroll waited for a second, then smiled and said, "Gotcha!" laughing his ass off all the way back to the pits.

Anticipation of the first production Vipers was intense, not just with the press and enthusiasts, but celebrities as well. Jay Leno and Tim Allen wanted Vipers, and Tom Kowaleski arranged for each of them to drive a Viper for a day before the long-lead press event. Winkles was lucky enough to draw the longest straw and chaperone Leno. Dick picked Jay up, and asked if he would mind stopping by Shelby's house to drop off some posters for Carroll to autograph.

"You know Carroll?" asked Jay excitedly, "I've never met him." Leno eventually found his way to Shelby's house and Carroll welcomed them in for refreshments and a tour, including a new, unassembled Cobra in the basement. After a nice visit, Jay asked Dick to take a photo with him and Carroll and send it to him.

The 1991 long-lead Press event started with dinner at Spago's in Beverly Hills, a night at the Plaza Hotel, followed by driving the Vipers to Tehachapi for an overnight stay, before track testing at Willow Springs and returning to Beverly Hills. The reviews were fantastic. Viper was doing its job: creating interest in a revitalized Chrysler!

The first Vipers arrived at dealers in May 1992, and Shelby continued to act as a spokesperson for the Viper's performance. It was clear, however, that Chrysler and Shelby were starting to drift apart. As Lutz' relationship with Iacocca deteriorated, so did Carroll's involvement with Chrysler. Whether real or imagined, there were suspicions that Carroll was just too close to Lee. It probably didn't help that Carroll let it be known that he thought the Viper too heavy, and would have preferred a lighter aluminum V-8. When Iacocca retired at the end of 1992, Carroll's association with Chrysler soon ended.

Leno tests the Viper for a day with Winkles. (Courtesy Dick Winkles)

Jay Leno and Carroll. (Courtesy Dick Winkles)

THE LEGEND AND A NEW SNAKE

Lutz and the Viper GTS Indy pace car. (Courtesy FCA)

In the summer of 1992, John Fernandez and I attended Shelby's induction ceremony into the Automotive Hall of Fame. Carroll was there with his new wife, Lena. John, being very close to Shelby, quietly inquired, "Why'd you get married again?" Shelby jokingly replied, "I need somebody to give me my pills!" After the ceremony, Carroll came over and told me he was thinking of doing another sport car, and wanted to know if we might have a suitable engine. All we had was the iron block Magnum V-8, which I did not think suitable, nor did I think that politically such a deal would be approved. I never did learn whether Shelby intended to revive the car he had started with Fernandez, or if this was his first foray into what was to become the Series 1.

Shelby's last official appearance for Chrysler was with François Castaing, at the 1993 LA press conference, for the introduction of the Viper GTS coupe. Although Shelby's involvement with Viper ended, significant events involving Fernandez, Hannemann, and I would end up tying Shelby's past triumphs with future Shelby adventures.

If the RT/10 was inspired by the 427 Cobra, then the GTS owed its lineage to the Shelby Daytona, designed by Peter Brock. First shown in LA, the GTS was the result of a rendering done by Bob Hubbach, at the request of Tom Gale. As a courtesy, Tom previewed the sketches with Brock prior to the GTS concept reveal. Viper would again pace the Indy 500 with the GTS in 1996 – only this time with Bob Lutz at the wheel. About this time, I moved from the minivan to the small car platform team (to which the Viper team reported). The international long-lead introduction of the GTS in Europe was an unforgettable Kowaleski extravaganza, starting with a parade down the Avenue des Champs-Élysées, then drives at Rheims (site of Shelby's first GP race), Spa and the Nürburgring hosted by world champion Phil Hill and Brian Redman. That story is for another time, along with the many other adventures that only Viper team members can describe.

Earlier that year François brought up the potential for a next generation mid-engine Viper. Personally, I thought this was a great idea: Viper and GTS paid homage to the Cobra and Daytona, why not the do same with the legendary mid-engine Ford GT40? Roy Sjoberg kicked off a study to see how much of the existing Viper componentry could be salvaged, using either a reverse engine layout or conventional transaxle. Roy assigned Jim Sayen to lead a small team on the GTM study ('M' for mid-engine) at the third Viper Owner's Invitational held at the Indianapolis Speedway that May. On October 17, the team was scheduled to present the two bucks and rolling mock-up to Castaing, Lutz and Gale, along with competitive vehicles. For some reason I could not attend the meeting. After the meeting, François sent me a congratulatory note to share with the team, "you did a great job," but later told me that Bob did not support the idea – probably because of the investment required. We put the study on ice.

With the demise of the Lamborghini F1 program in late 1993, and the successful launch of the Viper, the 'Motorsport Committee' was searching for something to enhance Chrysler's performance image. They decided to develop the Viper for GT1 and GT2 racing in Europe and the US, and put Neil Hannemann in charge of leading the development, thanks to his combination of engineering and driving skills. Bob Lutz and François Castaing introduced the GTS-R factory built racer at Pebble Beach in August of 1995. Designed for independent race teams, it was priced at $200,000 and homologated for ACO, IMSA and FIA sanctioned events. Journalists were skeptical that the Viper could be competitive. Indeed, there were few factory orders, until Canaska Southwind Motorsport demonstrated its class-leading speed at the Rolex 24 at Daytona. To prove the competitiveness of the GTS-R in Europe, a factory effort was established, and François enlisted the help of his old friend, Hugues de Chaunac of Oreca, to run the team. Unlike the Shelby Daytona, it would take a little longer for the GTS-R to win in its class at Le Mans. For the 1996 race, the Viper was fast, but fragile. By 1997, the GTS-R was the FIA GT Champion, and finished first in class at Le Mans the following three years!

With the impending retirement of Roy Sjoberg, I decided to promote Fernandez (an unabashed racer, who had successfully created and campaigned the Neon ACR cars for SCCA events) to head the Viper team, and we began to think about what we would do for a third generation Viper. Fernandez and the team took one more crack at a mid-engine Viper, using the Ferrari F355 Berlinetta as a bench mark. Advanced Design explored several scale clay models before the project died once again. Fernandez went on to lead the Viper team in the development of the third generation, 2003 Viper. Viper's racing successes continued under John, with the first ever GT1 outright win at the Daytona 24 hour race, beating all of the GTP cars. The Viper SRT-10 ACR package developed for SCCA racing went on to become one of the best GT1 cars in SCCA history.

Things were starting to change at Chrysler. Lutz' position changed from President to Vice-Chairman with CEO Bob Eaton taking over presidential responsibilities. François retired. Lutz later retired. Then came Eaton's disastrous 'merger of equals' with Daimler-Benz. My partner in manufacturing, Shamel Rushwin, secretly told me he planned to leave for Ford and encouraged me to do so, as well. I stubbornly clung to the idea of staying, but on the way home from the first joint DaimlerChrysler management meeting in Seville, Spain, I realized that Camelot had been lost.

Time to move on to the next chapter.

Chapter 3
THE COURTSHIP OF CARROLL AND FORD

March 1, 1999, I dressed and left for work, but instead of turning left towards Chrysler's Tech Center in Auburn Hills, Michigan, I headed south towards Ford's 'Glass House' in Dearborn. In the meantime, my partner in the DaimlerChrysler 'Great Escape,' Shamel Rushwin, was heading south from his home in Rochester Hills. Coming upon Ford headquarters, I took the ramp down to the executive garage, where security whisked me up the executive elevator to the 11th floor. Newly appointed CEO Jacques Nasser greeted me at the elevator. Jac grabbed my arm, led me up the circular staircase, past the life-size portraits of Edsel and Henry Ford, to the 12th floor. We headed to the southeast corner office for a brief introduction to Bill Ford Jr. Bill welcomed me, and then Jac proceeded to guide me to every office on the 12th, and then 11th, floors to personally introduce me to the other corporate officers. We stopped at the northeast corner of the 11th floor, next to the 'Thunderbird Room.' This was to be my new home for the next phase of my automotive career. By the next morning the *Wall Street Journal* headline in *Who's News* read: "Two Top DaimlerChrysler Executives Leave Company to Join Ford Motor."

I spent the next week getting to know the other executives, my staff, the facilities and product plans. There were numerous executive changes when Bill Ford became Chairman and Jac Nasser CEO. Consequently, Ford instituted a six-week 'shadowing' program for new recruits to spend time getting to know the executives they would be working with. My assignment for the second week was to shadow Jac. As luck would have it, Ford would be closing on the purchase of Volvo in Sweden. I packed my bags and headed to the Ford terminal where the Gulfstream was waiting for Ford's executive team. Since the closing was largely ceremonial, the party started early on the flight over. The discussion among the execs was about the business, and, in particular, how Ferdinand Piëch was amassing and positioning a host of brands at VW. As the red wine flowed, and still smarting from failing to get the mid-engine Viper approved at DaimlerChrysler, I took the opportunity to pitch the idea of doing a modern version of the famous Ford GT40 that had won Le Mans four years running. Richard Parry-Jones, global head of Product Development, picked up on the idea and began talking it up to Jac and the other execs. By the time we arrived in Sweden, they agreed to kick off a skunkworks project to explore the reincarnation of the GT40.

The next day was full of never-ending formalities and celebrations as AB Volvo sold the Volvo car division to Ford Motor Company, culminating in a lavish Scandinavian dinner, with speeches about AB Volvo entrusting their 'bride' to the new 'groom.' The toasts then began, each started with a different flavor of Swedish schnapps, a drinking song, downing of a shot, and then another toast and song, downed with another shot. These proceedings carried on long into the night. Not being much of a drinker, I tried to 'short' my shots, but nevertheless felt ill the next morning. Needless to say, the flight back to Dearborn was subdued as we all recovered from the revelry.

To my chagrin, when we got back to Dearborn, Richard Parry-Jones assigned the GT40 project to Neil Ressler, head of Ford Research. Neil pulled together a small team in the basement of the Product Development Center. The team included designer Camilo Pardo from J Mays' design staff, and Fred Goodnow, a talented engineer who had worked on every ill-fated mid-engine supercar project at Ford (Fred had joined Ford from AMC where he developed the AMX/3 with Dick Teague and Bizzarrini). Most SVT programs had aggressive code names like Condor or Terminator, but Neil chose the name 'Petunia' so as not to raise suspicion. Over the years, Neil and I became fast friends and he graciously invited me to the skunkworks for regular reviews of Petunia. While the team was making progress, Petunia never looked quite right: tall, narrow, with the requisite short overhangs and tight rectilinear surfaces that were in fashion. It did not have the strength or sex appeal of the GT40 – sort of like a distant offspring. The project seemed stuck in basement purgatory, despite backing from top management.

Early clay model of Petunia. (Courtesy Larry Edsall)

Meantime, as Vice President of North American Large and Luxury Cars, it was time for me to focus on mainstream commercial passenger cars. When I went to the cupboard, however, I discovered that the shelves were bare. Other than the 2000 Focus that we were about to launch (*Motor Trend*'s 2000 Car of the Year), and a face-lift of the 1986 'ovoid' Taurus, there was little in the product pipeline. Even the Thunderbird, scheduled for a 2002 introduction, had grossly overestimated production volumes. I had to inform Bob Rewey, Global EVP of Sales and Marketing, that we needed to scale back production capacity from 50,000 to 25,000 units a year. The investment budget for passenger car programs had been drastically cut to support the growing and profitable truck and SUV product lines. Jac Nasser assigned me to lead a 'capstone project' to put together a recovery plan. This team included executives from each of the functional groups: finance, sales, manufacturing, etc. The Ford management team enthusiastically approved of the final presentation. As time rolled on, however, programs were either cut or deferred, as the economy started to plateau, and trucks/SUVs absorbed more resources. The biggest blow, however, came when the Firestone tire crisis hit the Ford Explorer, putting the reputation, finances and very existence of Ford Motor Company in jeopardy.

Fortunately, a new Mustang was in the product plan for 2004, as the current Mustang, based on a Ford Fox platform dating back to the 1978 Fairmont, was clearly obsolete! Early on, design head J Mays (who had joined Ford from Audi a year earlier) and I were walking around the styling dome reviewing potential design themes for the new Mustang. J asked if I knew Carroll Shelby, and if I could introduce him. "Funny you should ask. If we are going to do a new Mustang, I think we need to plan on a Shelby GT350 and GT500 as well."

As I started preparing for a meeting with Carroll, I encountered resistance. Didn't I know about the feud between Carroll and Edsel Ford II? They had not talked to each other in years! Ford legal advised me to stay away, as there was an uneasy trademark truce between Ford Motor Company and Shelby. If I recall correctly, the Cobra trademark could be used on Mustangs by Ford, and Cobras by Shelby. Ford could use GT500, but GT350 was still in dispute. Then there were those who resented that Iacocca had recruited Carroll to Chrysler. Moreover, it seemed the myth of the feud was larger than life. No-one wanted to touch the subject.

I was aware of many conflicting stories about the dispute between Edsel and Carroll. They had been close for many years. Back in the mid-60s, Henry Ford II ('The Deuce'), had sent young Edsel to intern at Carroll's shop in California. Shelby took Edsel under his wing, and had him start at the bottom, washing parts. Edsel witnessed first-hand the racing success of the Shelby/Ford collaboration, culminating in the 1-2-3 finish of the GT40s at Le Mans in 1966. The indoctrination of Edsel took hold, and he went on to become one of the few 'car guys' and racing aficionados among Detroit's Big Three executives.

As Edsel moved up through the ranks at Ford, one stop was as marketing plans manager at Ford Division. To help move the metal, Edsel created a 1984 Mustang Anniversary Edition GT350. Shelby's lawyers sued for trademark infringement. Edsel could not help but be angry and took it personally, "He sued me!"

Years later, I asked Carroll what really happened. "Aw, I was sick in the hospital, and my lawyers sued." I doubt Carroll's account was entirely accurate. Shelby had learned early in life the value of a brand and its trademarks, from his famous striped bib 'racing' overalls, to his bolo hat and trademarks of GT350, GT500 and KR: 'King of the Road.' More likely, Carroll approved the suit while ill, not fully thinking through the implications of his actions.

To the outside world of car enthusiasts, it was only natural that Ford and Shelby reunite. I forged ahead,

THE COURTSHIP OF CARROLL AND FORD

and sought the support of others. I talked to Keith Crain, the *Automotive News* newspaper mogul, and learned, to my surprise, that he had been pushing Ford to bring Shelby back. Years later, I learned from John Coletti that efforts to bring Carroll back began as early as 1995. In working with the Mustang Club of America to explore the feasibility of a Mustang Museum, it was apparent to Coletti that such a museum would be incomplete without the names of Lee Iacocca and Carroll Shelby. In 1997, John Clinard (Ford's western region PR manager) set up a meeting with Shelby for Coletti to gauge Carroll's interest in returning to the Ford fold. Carroll was interested, but recognized the issue Ford might have, given their raucous past. Coletti then went to Edsel to get his recommendation. While Edsel would consider involving Iacocca, he was not ready to make peace with Carroll. Keith Crain and Heinz Prechter (the charismatic founder of ASC, American Sunroof) were on the museum committee advisory board, and John asked them to quietly push for reconciliation. Keith Crain, having been independently asked by both Coletti and myself, would continue in the intervening years to lobby both Edsel and Carroll, as he was close friends with both. Finally, Shelby asked former editor-in-chief of *Car and Driver*, William Jeanes, also a friend and neighbor of Edsel's, to let Edsel know that Carroll 'wanted back in to Ford.' Edsel was amenable, and in the spring of 2001, William arranged a meeting at the Grosse Pointe Club, aka the 'Little Club' (reportedly funded by Horace Dodge after he was thrown out of the 'Big Club' due to his boisterous behavior). Jeanes brought the two together on 'The Terrace' for brunch, and left them alone to patch up the relationship. From afar, William watched them conclude the meeting with a handshake. Now it was time to figure out how to signal that Carroll was back with the Ford family.

The breakthrough officially became public at the Pebble Beach Concours d'Elegance in August 2001, at the suggestion of John Clinard. Back in 1988, Edsel instituted the Spirit of Ford Award to recognise individuals who made outstanding contributions to Ford racing. Notably absent from the list of recipients was Carroll Shelby. John did not think this was right. After much thought he phoned Edsel and said, "After we're all dead and gone, the world will wonder why Carroll wasn't recognized." Edsel responded, "You're right. I'll come and present it myself." Clinard invited Carroll to the Ford dinner party Saturday night as a special guest, with no mention of the award.

Ford PR rented a beautiful beachfront house, overlooking the golf course and the Pacific Ocean, to host a lavish evening party for the press to meet legendary drivers who had piloted Ford race cars over the years: Jackie Stewart, Phil Hill, Bob Bondurant, Dan Gurney, Parnelli Jones, Sir Stirling Moss and, of course, Carroll Shelby, among others. In attendance were Jac Nasser, Edsel Ford, J Mays, myself and most of the executive management team, along with Keith Crain and other 'A list' journalists. Carroll arrived late with his new wife Cleo. As Edsel and Carroll met, they hugged and signaled that any bad memories were forgiven and forgotten. It happened to be Cleo's birthday, so we all celebrated, singing Happy Birthday, as a cake lit with sparklers arrived. Afterwards, the gentlemen gathered around the fire, trading racing stories and barbs over beers and liquor, while the ladies retreated to a nearby tent overlooking the ocean. Evi Gurney and the other women entertained each other recounting the exploits of Carroll, Dan, and the other racing greats.

After dinner, Jason Vines (VP of Public Relations) invited Edsel to address the group. Edsel began by noting that 2001 was a significant milestone: "100 years of Ford Racing – the 100th anniversary of my great-grandfather's one and only race." Edsel then went on to note Carroll's indispensable role in the history of Ford Racing, and shared a couple of personal vignettes:

"My father ... called Carroll Shelby out in California and made some arrangements. I flew out on American Airlines, took a cab, and pulled up to Carroll Shelby's house. I rang the doorbell and there was this absolutely

Edsel Ford presenting Carroll Shelby with the Spirit of Ford Award, August 18, 2001. (Courtesy John Coletti)

stunning woman who answered the door. She said, 'I'm Mr Shelby's housekeeper.' Well, even at the young age of 17, I could hardly stop from smiling."

"My second and most vivid memory was Shelby American's GT40 all-American victory at Le Mans in 1967 with Dan Gurney and A J Foyt behind the wheel."

Edsel then presented the Spirit of Ford Award to Carroll, who responded by telling his own story about Edsel. "Three times a day, Henry Ford II would call me to make sure things were okay. He'd tell me, 'You make damn sure he washes those parts good.' Like I was going to tell him he wasn't doing a good job?"

The ice had been broken, and we started to make plans for Shelby's involvement. As word of the make-up between Edsel and Carroll made its way through the organization, the barriers started to break down. The next step was to meet with Carroll in Las Vegas during the SEMA (Specialty Equipment Market Association) Show, where manufacturers, 6000 suppliers, 70,000 buyers and untold numbers of enthusiasts gather to display new products and hundreds of custom cars. I was to be a keynote speaker at the opening lunch. John Coletti, Director of SVT, and Mike Zevalkink, Director of the (newly formed) Ford Performance Group, joined me. While I was at SEMA, Coletti went to Carroll's house to work out an agreement for Shelby appearances at Ford events, while Ford would supply what Carroll called "soft support" (ie technical data) so Shelby American could build performance Ford products under the Shelby brand.

As an aside, in 2000, few knew that CEO Jac Nasser authorized John Coletti to approach Carroll regarding the sale of Shelby American for $20M, or that John had already reached tentative agreement to supply Ford modular engines for the Shelby Series 1 in place of the Oldsmobile motors. The Series 1 project had come about while Carroll was suffering kidney failure, but Don Rager convinced him to proceed with building a new Cobra at a new Shelby American facility adjacent to the Las Vegas Motor Speedway. John Rock, who had taken over Oldsmobile, supported the project by agreeing to fund $1M for two prototypes, provide engineering assistance, and supply Olds engines and GM components at transfer prices. Things started to fall apart when Rock retired because of GM's hiring of Ron Zarella to implement "brand management" – an epic failure. The agreement for $1M, engine supply and transfer pricing disappeared. Carroll needed a replacement for the Oldsmobile engines, and let John drive his first Series 1 prototype at the Las Vegas motor speedway. Although the prototype had many issues and shortcomings, it had potential as a modern-day roadster. John met with Neil Ressler and recommended that Ford consider the acquisition of Shelby-American to expand its portfolio of brands. Upon receiving the go-ahead from Nasser, George Joseph and Coletti went to Vegas to meet with Shelby and his new business partner, Larry Winjet. Timing is everything in this world, as Carroll had previously sought $5M in financial backing from Winjet, owner of Venture Industries, to keep the Series 1 program going. At the meeting, Winjet smelled deep pockets, valued the company at $100M, and demanded a 50% stake. The negotiations abruptly ended. After the meeting, an embarrassed Shelby apologized to John, as he had no choice in the matter. Not long after, Venture Industries went bankrupt, and Carroll bought the remains of unfinished Series 1s at auction. One can only wonder what might have happened had Coletti and Shelby been able to get together before Winjet's involvement.

Since the product cupboard was bare for the near term, we had to focus on creating specialty models to keep Ford cars fresh in the minds of the consumer. The 2001 Mustang Bullitt had been first, which we introduced with Chad McQueen recreating the chase scene in the streets of San Francisco, followed by the 2002 SVT Focus. We knew we needed to do much more, and Shelby was just the guy to help do it. Shelby's first official engagement was at Ford's 100 Years of Racing Festival at Greenfield Village in October 2001. I remember watching Edsel drive the Ford 1901 'Sweepstakes' race car in the rain to recreate Ford's victory over Alexander Winton in a $1000 sweepstakes race.

The matchmaking was complete. Little did I know it would take another five years to consummate the marriage with a Shelby badge on a Ford production car! Each step of the ensuing journey involved convoluted twists and turns.

Chapter 4
PROJECT PETUNIA

The courtship had begun, but it took until 2002 for the relationship to take hold. Fortunately, Petunia, aka GT40, aka Ford GT, provided the catalyst. Still frustrated with the lack of traction in moving project Petunia forward, and still unhappy with the look of Petunia, I kept sketching images of the original GT40 while sitting in the endless executive meetings that were steeped in corporate bureaucracy. It was in one of those meetings that Richard Parry-Jones, while lecturing to the Product Development staff, caught me sketching. Grabbing the sketch from my table, he held it up for the audience to see and handed it to J Mays, saying, "We've got to get serious about this."

J Mays had moved Petunia from the basement of the Product Development Center to a studio at the Schaefer Road facility. Apparently, John Coletti was not happy with the looks either, and told Neil Ressler. Neil spoke to Jac Nasser and scheduled a design review, while Coletti arranged to bring an original Ford GT40 press vehicle over from Europe to provide inspiration. At the meeting, J presented some sketches by Camilo Pardo and Kunihisa Ito, touting modern interpretations of the Ford GT. Never a shrinking violet, Coletti said, "That doesn't look anything like a GT40." At which point Jac said, "I've only got a few minutes, let's make this simple. I want this concept to look like that car" and pointed at the GT40.

Jac left and Camilo asked J, "So what do we do now?"

J said, "I don't give a damn," and walked out.

Coletti looked at Camilo and ordered, "You heard the man, get with the program!"

Eventually, J bought into the new direction. In a meeting not long after, I noticed J Mays sketching a new GT40 profile, much as I had imagined. He leaned over and whispered, "I get it. It's the proportions: long front overhang, short tail and sexy curves." Rather than come up with a new GT40, we would simply "reissue" the legend, as J put it. In March 2001, they moved Camilo into the 'Living Legends' design studio. As the new GT40 took shape in clay, with an original perched next to it for inspiration, we all fell in love with Camilo's interpretation, a tightened up and improved riff on the original.

Now that we had a design we liked, J and I decided it was time to get Petunia out of purgatory.

"Screw it!" said J. "Let's make it the star of the 2002 Detroit North American International Auto Show." I readily agreed. If the GT40 was a hit with the press and public, how could we not put it into production? Fortunately, the Petunia skunkworks, led by Goodnow, had been studying how to make Petunia feasible. Working with supplier Multimatic, they developed a space frame chassis, while Coletti's SVT group was developing the supercharged, four-valve MOD V-8s. Unfortunately, the original Petunia design had decidedly different dimensions than the new GT40 clay. I called Coletti and Goodnow into the Legends studio and told them we needed to put their chassis, suspension, and powertrain into Camilo's clay, and have a running show car by January! Despite some grumbling about the low roof and package changes that would be required, Fred picked up the gauntlet and ran with it! I used to think of Fred as a sparkplug, but Coletti nailed it: Goodnow was a Gila monster – once he latched onto something, he never let go!

The skunkworks had been looking at different methods for manufacturing Petunia. One of the alternatives was to have the GT40 built by an outside supplier. Pininfarina, wishing to fill its niche vehicle plant, had made a proposal to manufacture Petunia. I first met Andrea Pininfarina during aerodynamic testing of the Fiat Spyder Turbo in the early '80s, while Andrea was interning for his father at its wind tunnel in Grugliasco, Italy. I preferred that Ford build the GT40 in-house, but it was more important to explore every manufacturing option. Better to find a way, than to not build the GT40 at all. We planned to meet during the Frankfurt Auto Show at the Pininfarina display stand. I met with Andrea and his engineering team on September 11 to review their proposal. As they finished, I thanked them, but made Andrea aware that, if possible, we would build the car locally with Ford employees. Returning to the Ford stand, I stopped into the pressroom, only to see the second plane strike the south tower of the World Trade Center. Time stood still and the world stopped as the tragedy unfolded before everyone's eyes. Thoughts about the GT40 vanished, as the most important things in life came into sharp focus.

It took time for the world to recover, but humanity always pushes on, and so did the team. Only three months remained to complete the show car, with much left to do. John Coletti and I invited Carroll to see our secret project. As the leader of the 1966 GT40 Le Mans winning race team, it was only natural to include Carroll in the program. Meantime, J wanted to make sure the reveal at the Detroit auto show would be perfect: he asked the Ford display staff to create an introductory film in the '60s style of the classic John Frankenheimer film *Grand Prix* mixed with a touch of Steve McQueen's *Le Mans*. The GT40 concept made its debut on January 12, 2002. Nick Scheele, the new President and COO of Ford introduced the film clip by saying that when you talk about the "heritage to propel Ford into the future … there is no better product to sum it all up" than what they were about to see. The film opened in the dark with a radio announcer describing Ford's 1-2-3 finish at Le Mans in 1966, quickly moved into a spit-screen montage of Le Mans racing clips including Henry Ford II and Dan Gurney, as a live orchestra added to the mood. Soon, the film transitioned into flashes of the new GT40, followed by classic GT40s rumbling around the stage as the yellow GT40 concept drove onto the turntable. Out popped Bill Ford doing what every Ford GT owner does at least once: hit his head on the wrap-over door! The crowd of journalists cheered as Bill announced the return of the GT40, and then introduced other Living Legends: "Walking Ford Ovals" Phil Hill, Lloyd Ruby, Bob Bondurant, Jackie Stewart, and, of course, Carroll Shelby.

Reaction by the media and the public at the show was outstanding. One might have assumed that this

Bill Ford introducing the Ford GT40 Concept.
(Courtesy Ford Motor Company)

would be enough to get the program approved. There were other factors in play, however, that helped ensure that the GT40 did not suffer the same fate as prior mid-engine supercar programs. On October 30, 2001, Jac Nasser was out as CEO, and Chairman Bill Ford assumed the title of CEO as well as Chairman. Nick Scheele called me and J Mays up to his office. As I recall, Nick said we needed a big announcement to demonstrate that Bill was firmly in charge, leading Ford into the future as it approached its centennial. Should we announce the GT40, or the "Forty-Niner" as Nick called it? Ford had introduced the Forty-Nine concept car at the 2001 Auto Show to enthusiastic reviews. Trying to conceal my emotion, I opined that we could get the GT40 into production as a limited volume vehicle sooner than a higher volume Forty-Nine, even though I would have loved to do both. Nick seemed to favor the "Forty-Niner," but asked us to think about it and come back to him. Ever the salesman, J hatched up a plan to convince Nick. On our return visit, J asked Nick to come down the executive elevator to the garage. As the doors opened, there stood the GT40 painted in the Gulf blue and orange livery, its bright headlights pointing straight at Nick. "Oh you bugger!" he cried, nearly falling to his knees with flashing childhood memories of Le Mans. We had our decision, and Scheele asked us to come back with a business plan.

Coletti and Goodnow got the great news, and put together the plan. They presented a two-page summary to Nick. It was probably the shortest program approval in Ford history, but it came with a caveat: Bill wanted three "production representative" cars completed for the Ford centennial celebration that was just 15 months away! In actuality, Bill only asked for one vehicle, however Ressler was concerned that some executive might wreck it, and asked Goodnow to build three cars. Somewhere along the journey, they became the red, white and blue Ford centennial cars. Ever ready for a challenge, Fred went off and drafted a thick manifesto defining what must be done differently to achieve the objective. We all bought in, and the starting gun sounded!

Six weeks after the unveiling, Bill Ford announced that the GT40 would go into production. The press and automotive enthusiasts applauded the announcement, albeit with a bit of skepticism. To some extent, the skepticism was justified considering the history of past attempts, and in fact, there was no GT40 team in place, just Goodnow's manifesto and the feasibility study. Nick Scheele had promoted me to VP of Ford North American Product Development to combine car and truck engineering operations. "I think you're going to need some help with the GT40. Would you mind if I asked Neil Ressler to come out of retirement and consult as a technical advisor?"

"We'll take all the help we can get!" I replied, knowing that Neil would help the team hit the ground running. Nick also suggested that we contact his friend, Tony Johnson, who owned Saleen – perhaps there were some synergies. While we found no synergies with Saleen's S7 supercar, there was a bit of serendipity, in that Neil Hannemann from the Viper GTS-R racing program was working there, and had driven and worked for Carroll Shelby years before. I had also insisted that the GT40 have an aluminum body and frame, and suggested that Fred contact Mayflower Ltd, since they had built the aluminum Plymouth Prowler bodies for Chrysler (Tony Johnson later purchased Mayflower, after divesting Saleen).

Things were quickly starting to come together for both team management and key supplier selection. Goodnow spent the next 48 hours creating job descriptions. Next Ressler, Coletti, Zevalkink, and Goodnow interviewed and selected the best of the best within Ford. By March, we called the GT40 team together for a kick-off meeting. Goodnow, Ressler, Coletti, and I addressed the team, told them it was a once in a lifetime opportunity, and what the goals were, while Goodnow emphasized the commitment that would be required (no vacations, no babies!). I told the team to imagine pulling up to a stoplight on Woodward. To their left would be an exotic red car badged with prancing horse, and I asked them to imagine saying to themselves: "My car is faster, better looking, handles like it is on rails, and I didn't get ripped off!" To bring it all home, we asked Carroll to meet with the team, explain the heritage and inspire them to create another legend.

To satisfy the skeptics, we announced the 'Dream Team' to the press at a dinner held in the Cranbrook courtyard, a private school founded by famed architect Eero Saarinen. The placemats introduced the GT40 Dream Team:

- **John Coletti**, director Ford Special Vehicle Team programs and GT40.
- **Fred Goodnow**, GT40 vehicle engineering manager and former SVT programs manager.
- **Jamal Hameedi**, GT40 program manager and former Ford Powertrain and Off-Road racing manager.

- **Neil Hannemann**, GT40 chief engineer and former chief engineer of Saleen S7 and former program manager for the Dodge Viper GTS-R.
- **Camilo Pardo**, GT40 chief designer and head of design for Ford's SVT products.
- **Tom Reichenbach**, GT40 vehicle development manager and former dynamics manager for Ford's racing programs, including Formula 1, CART and NASCAR.
- **Neil Ressler**, former Ford vice president of advanced engineering and chief technical officer, now GT40 senior technical advisor.
- **Carroll Shelby**, owner Shelby American Inc and senior technical advisor to the GT40.

Also present were Jack Roush, Tony Johnson, Gordon Boyd, and Kenneth Way representing our major supplier partners:

- **Roush Industries** – Powertrain development.
- **Saleen Inc** – Niche manufacturing system.
- **Mayflower Ltd** – Chassis and body systems.
- **Lear Inc** – Interior systems.

That evening made it clear to the pundits that we were serious! Next on Goodnow's agenda was getting the GT40 workhorse vehicles designed, built, and on test. Colocated in 'Roush 56,' the team would review progress in the 'Le Mans' conference room connected to the garage area dubbed 'The Pit,' where Ford engineers, designers, and supplier partners worked together to make their dream assignment a reality. On November 12, 2002, Neil Ressler and I drove GT40 'Workhorse Number 1'– now renamed the Ford GT for reasons I will explain later – out of The Pit to the cheers of the team and the press.

I think most team members thought Shelby's role would be limited to that of 'spiritual advisor,' but Carroll took the role of 'senior technical advisor' seriously. He could still expertly handle a car and spent a lot of time evaluating the GT40 as development progressed, starting with the workhorse. It was no accident that workhorse 1 was "left" in Las Vegas after testing was completed. It now resides among other historic Shelby Fords at the Shelby Heritage Center!

Workhorse Number 1 breaks cover with Ressler and Theodore. (Courtesy Ford Motor Company)

PROJECT PETUNIA

Shelby and Hannemann discussing the Ford GT workhorse.
(Courtesy Larry Edsall)

Work continued at Roush 56 in Dearborn, and the new Saleen facility in Troy, at a feverish pace towards the centennial deadline. As 2003 rolled around, the build of cars one, two and three began at the Saleen facility. Fred was recovering from back surgery, but was still buzzing around the shop on an electric scooter barking out orders. I had surgery as well, but would sneak over to check on progress whenever I could. Assembly started in March, using the first parts built off prototype or production tools. They had to work perfectly first time, and, as a testament to the team and the suppliers, most of them did! As June 16 approached, the red, white, and blue cars were completed just in time for a photo shoot at the glass house. The building was buzzing as word spread, and those that could, snuck into the studio for a peek. The cars looked awesome!

The opening day of Ford's centennial celebration was cold and wet. Held at the 52-acre site of World Headquarters, cars and fans arrived from all over the country to experience a carnival-like atmosphere. Bill Ford led the opening parade through the packed crowd in the new F-150, followed by the Ford GTs: White #1 with Jim O'Connor (VP of Sales), myself driving red #2, and Coletti in blue car #3. The evening centennial celebration opened with the red, white, and blue Ford GTs roaring up the ramp and onto the stage. Scheele popped out of the white car to introduce Jackie Stewart, while Shelby, Coletti, Gurney, and myself exited the other two GTs. The crowd cheered. Little did they know, we were the opening act for Beyoncé Knowles, in her first solo appearance after breaking off from Destiny's Child! We celebrated and watched her perform from the Ford tent, while Edsel's young boys explored the GTs, and learned the significance of their history.

After the introduction at the centennial, it was time to formally announce the engagement of Ford and Shelby at the Pebble Beach Concours d 'Elegance. After breakfast at the Lodge's Stillwater Restaurant overlooking the 18th hole, the press watched as I drove up in the red Ford GT and announced that now that we had reintroduced one legend, it was time to introduce another. Carroll Shelby roared up onto the temporary stage in a deep blue 427 Cobra. We revealed that Ford and Shelby were joining forces again to develop Shelby performance cars and parts. At the conclusion of the reveal, a media scrum surrounded Carroll, shouting out questions and taking photos. Eventually, the press and most of the Ford team left for other events, leaving Carroll alone with Cleo.

As the last straggler departed, I overheard Cleo, who, like a protective mother, whispered: "They don't appreciate you."

"Quiet," Shelby replied, "we're only just beginning." Pretending not to hear, I walked up and invited them back to the Ford suite. Meantime, the Ford press office was frantic. The phone lines lit up like Christmas trees with calls from all over the country. Manning the phone lines, I spent the rest of the day tap dancing about the future products that

Red, white and blue Ford Centennial GTs. (Courtesy Ford)

Ford Centennial: Dan Gurney, John Coletti, Chris Theodore, Nick Scheele, Carroll Shelby, Jackie Stewart. (Courtesy *Motor Trend*)

PROJECT PETUNIA

August 15, 2003 Pebble Beach announcement: Ford and Shelby Join Forces! (Courtesy Ford Motor Company)

1965 Ford advertisement for the 'Ford GT' with Ken Miles and Carroll Shelby. (Courtesy Ford Motor Company)

were to come. The cupboard was bare, but it was time to start filling it.

By now, some might be wondering how the GT40 became the Ford GT. It turned out that Safir GT40 Spares Ltd claimed to own the rights to the name. Ford had secured the naming rights for the concept car, but the negotiations for a production car were not secured. Ford authorized the CFO to offer up to $2M and two production GT40s. For some reason, the negotiations ended abruptly. In the end I was happy to change the name for three reasons: 1. Had it been named GT40, the 'Ford' would have been silent – changing it to Ford GT reinforced the desired halo effect; 2. The initial Mark I, created by Roy Lunn, had been called the Ford GT – GT40 started out as a nickname because of the car's low, 40 inch height; and 3. Mike Zevalkink learned that Holman-Moody thought they owned the GT40 naming rights, so a potential legal conflict was avoided.

In April of 2004, we received a letter from Roy Lunn, 'Godfather of the Ford GT,' who wrote to make sure we knew the facts "before some published distortions became reality." I had the good fortune of knowing Roy when our careers briefly overlapped at American Motors, where he engineered the unibody Jeep Cherokee and Renault Sport Cup spec racers. Roy was also responsible for the Mustang I, a mid-engine car I greatly admired, about which he wrote an SAE paper that I keep a copy of to this day. Lunn's ten-page letter is too long to include here, but I have appended a transcript for those who want to know the history straight from the horse's mouth. I passed a copy of the letter on to Carroll, and he confirmed that Roy's account was accurate. The only thing Carroll added was that Eric Broadley, of Lola fame, left the program a year later because the big company environment "was driving him crazy." Roy was inducted into the Automotive Hall of Fame in 2016, 24 years after Carroll. Sadly, Lunn passed away a year later at the age of 92, but not before I obtained his autograph on a Ford GT poster.

With the Ford GT history and name settled, the team was racing towards the finish line. Shelby continued to be involved, evaluating the Ford GT's progress. Hannemann or Reichenbach would invite Carroll to test drive prototypes whenever they were on the west coast, or Carroll was in Detroit. The team rushed to complete the facilities for Job 1 and enter, as Elon Musk puts it, "manufacturing hell." I used to joke that the Ford GT came with frequent flyer miles: transaxles from the UK; aluminum panels from Riverside, California; bodies from Norwalk, Ohio; frames from Detroit; and engines from Romeo, Michigan. To boost troop morale, I asked Carroll to spend a day with the craftspeople building Ford GTs. We headed to Detroit to watch robots weld the aluminum space frames together. Inspectors cut every 13th frame apart, checking weld quality under a microscope. Carroll had never seen anything like it!

John Wyer, Eric Broadley and Roy Lunn near London. (Courtesy Ford Motor Company)

Next stop was the paint and assembly shop in Troy, Mi. Team members flocked to see Carroll, take pictures and obtain autographs. Shelby loved to take time with each individual, smiling, joking, and telling stories. Finally, it was off to the Wixom assembly plant, where they mated the engine and transaxle, installed it into the chassis, and fired up the Ford GT to run it on the chassis rolls before the final water-leak test. Carroll spent the rest of the day with the team, learning each step of the assembly process. By the time the day was done, you could sense the excitement of the team members. They knew they were part of the legacy of both Ford and Shelby.

Before we could start production, there had to be a sign-off trip of the final prototypes. Ressler, Zevalkink, Coletti, Goodnow, and I met the team in LA. Waiting for us were four Ford GTs, a Corvette, and a Ferrari 360. We drove through Death Valley, up to

Shelby touring the Ford GT assembly facilities.

Theodore and Shelby on the assembly line at Wixom.

Ford GT Sign-off Team Photo at Dante's Peak. (Courtesy Mike Luzaders)

Final Sign-off at Las Vegas Speedway. (Courtesy John Coletti)

Dante's Peak, and on to Las Vegas where we created as stir as we slow rolled down the strip. The next morning it was off to Shelby American and the Las Vegas Speedway, where we spent the day testing the cars on the handling track and oval. Shelby provided a little additional entertainment by having Gary Patterson bring out a supercharged Shelby Series 1 to drive. I stayed with the Ford GTs to make sure the boys didn't get overly aggressive (they did!), while Coletti went for a "quick trip" with Gary around Vegas in one of the Ford GTs. At the trip debrief, we decided it was 'all systems go.' The cars had performed flawlessly, with only a few minor fit, finish, and NVH issues to fix. Before he retired, John Coletti presented the program financials to Bill Ford, who summarized, "While the Ford GT Program was never intended to be a money maker, we did make a few bucks on the program." The entire Ford GT team created a great car, without needing the company to subsidize their effort ... they earned it!

There are too many other team highlights, memories and stories about the Ford GT to tell. Things like:
- Racing the Ford GT up 'the Hill' at the Goodwood Festival of Speed.
- Jay Leno auctioning the first production GT at Christie's for $500,000 plus commission.
- Press drives at Laguna Seca with Jackie Stewart and Dan Gurney, as they challenged each other.
- Leno taking the 'Zanardi shortcut' down the corkscrew, with me as passenger!
- The Ford GT recording a 211.89mph top speed at Nardo, Italy.
- Building a red Ford GT with a 'Gurney bubble' so Evi could surprise Dan on his 74th birthday.
- The European Press introduction on the track at the Petit Le Mans, etc.

Goodnow with Theodore's GT at Wixom end of line.

Hannemann, Theodore and Gurney at Laguna Seca. (Courtesy *Motor Trend*)

THE LAST SHELBY COBRA

Those stories are for another time. In 2003, you would have thought that Carroll, an octogenarian running on a spare heart and kidney, would have had enough excitement, but you would be wrong. A secret project, codenamed 'Daisy,' was in the works.

Jay Leno pitching the 1st Ford GT at a Christie's auction. (Courtesy Ford Motor Company)

Nair, Zevalkink, Coletti, Theodore and Ressler at Job #1.

European Ford GT press introduction at Le Mans. (Courtesy Ford Motor Company)

Chapter 5
CODENAME: DAISY

Every year in Detroit after the North American International Auto Show, J Mays and I would get together to plan concept vehicles for the next year's round of shows: LA, Detroit, Chicago, and New York. Mays' design team would provide suggestions, as did my Advanced Product Creation group. With the success of the Ford GT at the centennial, it was no surprise that a modern Shelby Cobra was at the top of both our lists. We also decided to do a new Bronco and Lincoln Mark X. Mays let Richard Hutting, manager of the Valencia Advanced Design Studio, know that we would be reviewing proposal sketches on our next trip out west – as J and I would pay regular visits to the Irvine and Valencia studios for design reviews. Now that the world knew about Petunia, we decided to call this project Daisy. We intended the codename to be a little tongue-in-cheek and sort of a tease. As J said, it would be "anything but a shrinking flower." Eventually everyone would know that we were up to something, but not know what. I called it a 'fan dance' – the most tantalizing secrets are the one that you know are there, but cannot quite see. In late March, we sent Manfred Rumpel to see Hutting, specifically to explore how to help with the packaging of the Ford GT suspension and a V-10 engine.

During one of my weekly program reviews with the SVT team, I mentioned to Coletti that I had kicked off Project Daisy.

"Have you talked to Carroll?" asked John, wanting to make sure he wouldn't react negatively when he found out.

"No, I forgot, but I know he'll love the idea." I asked John to invite Shelby to the first design review at the Hutting studio in April. Meantime, recognizing the historic significance of such a project, Mays had the presence of mind to cut a deal with TLC/Discovery for the first episode of a new TV show called *Rides*. They would document the historic development of Daisy. Bud Brutsman produced the show, and it was narrated by Jason Priestley.

On the appointed day, Coletti met Carroll at the Bel-Air Country Club. The valet pulled up in a blue 2003 SVT Cobra provided by Ford as a courtesy car. Coletti was shocked by the garish billet wheels – bright polished intertwining snakes – but did not say a thing to Carroll. Off they went up the I-5 freeway to Valencia, when a kid in a yellow Porsche pulled up alongside, inciting a race.

"This guy doesn't know who he is messing with," thought Coletti. Never one to turn down a challenge, Shelby set off, past 130mph, until the Porsche shut down. Carroll just smiled.

As they got out of the car at the studio, Carroll asked if he could put his hand on John's shoulder.

"Why?"

"Because I can't see so well."

John snapped back, "You were just doing over 130mph on the freeway, and now you tell me you can't see!" Out came that big Texas grin as I walked up to greet him and said "Nice wheels!"

"Aw, this real nice kid is a Shelby enthusiast and sent them to me, so I had to put them on the car and send him a picture." As we walked into the lobby, who else but our receptionist, Shelby, would be there to welcome him. The designers and clay modelers were standing at attention as he entered the studio, as if a major general was entering the room to review the troops.

Carroll put them 'at ease' with his charm and a few quips as he reviewed the wallboards full of exterior and interior design sketches. Shelby pointed out some sketches he especially liked, but complimented them all. Over on a round cocktail table was a beautifully detailed scale model of a chassis that Hutting had built himself. As we gravitated toward it, we noted that it had the traditional front engine and transmission, with the rear drive of the original Cobra, and cramped, skewed seating position.

Almost simultaneously, Carroll and I came up with the idea to use the Ford GT transaxle in the rear to improve both the interior package and weight

THE LAST SHELBY COBRA

J Mays and Chris Theodore discussing the chassis model with Hutting. (Courtesy Bud Brutsman)

distribution. We always planned to use the Ford GT suspension for Daisy, but at that moment, I had sort of an epiphany: With the Ford GT soon to go into production, we now had a collection of supercar components from which to develop Daisy and other high-performance cars. The excitement grew as we started to talk more about Daisy. This wouldn't just be a pretty concept car; this would be a "full-on runner," as J described it. I would have Hermann Salenbauch and Manfred Rumpel (manager of the APC chassis group and former chief chassis engineer on the Porsche 917 Can-Am car!) design a complete running chassis, so we could establish feasibility of a production car simultaneous with the development of the show car. The project took on a new sense of purpose when we really started leveraging the Ford GT.

Top right One of many Daisy CAD design proposals. (Courtesy Bud Brutsman)

Right Early model of Daisy. (Courtesy Bud Brutsman)

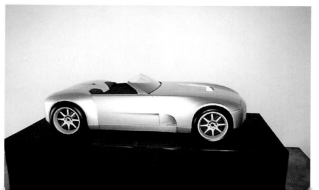

CODENAME: DAISY

So where did the V-10 come from? Back in the early days of the Ford GT, I called together guys from the Petunia skunkworks, SVT, Ford Racing, and Advanced Powertrain to discuss engine alternatives. There were four factions: One wanted a small displacement high-revving motor like a Ferrari V-8; another wanted twin turbos; a third pushed for a supercharged Mod V-8, like in the SVT Lightning; and I had suggested a large displacement V-10. I could see a horsepower war brewing, and a V-10 would allow us to supercharge or turbocharge it in the years ahead. The naturally-aspirated version would also be lighter and permit a lower center of gravity (Cg) than a supercharged V-8. The high-revving engine option was put to rest when I told the guys we needed at least 500 horsepower and 500lb/ft of torque. Turbocharging would have been a thermal challenge to execute in a mid-engine car in such a short time frame. Finally, Coletti convinced me that the only way Petunia could meet its timing objective was if we used the Lightning motors as surrogates in the Ford GT 'workhorse' vehicles.

Meanwhile, unbeknownst to me, the Advanced Powertrain team went off and built a V-10. Greg Coleman and Kevin Byrd called me one day and asked if I wanted to take a ride in a hot Mustang. Of course I did, not suspecting what was under the hood. Performance was phenomenal, but it was too late to disrupt the Ford GT program, even though the V-10 engine would have fit. Mays and I decided to put it in the 2003 '427' sedan concept – we fibbed a bit about displacement! I called Greg and said we needed another V-10. "Just write the check for the parts!" he replied. In the end, four V-10s were assembled and installed in the Mustang mule, the 427 concept car, Daisy, and a concept car planned for 2005. When all was said and done, the check ended up being more than half a million dollars!

Fortunately, the cameras were rolling as the brainstorming continued. Bud Brutsman asked Carroll to sit down and talk about the Cobra and Daisy for the camera. With his Texas twang and gravelly, but not deep, voice, Shelby began:

"I never dreamed it would become an icon, as it did. I was just trying to kick the shit out of Corvette …I'm 80 years old, transplanted heart and transplanted kidney. I wake up three times a night with new ideas. There are so many things left in the world that I want to do that I don't even care what I'm remembered as. I just want to be there to do a few of those things, and building a new Cobra is number 1."

John Clinard hosted a lunch for everyone at a

Shelby interview for *Rides*. (Courtesy Bud Brutsman)

restaurant near the Valencia studio where we had a fabulous time listening to Carroll's stories, beginning with the original Cobra. "I went to see Don Frey and Lee Iacocca about getting a Ford engine to do the car. I'm told that after the meeting Lee Iacocca said 'give that guy $25,000 before he bites somebody.'" As an amazing day ended, we agreed that Carroll would be involved in every step of Daisy's development. Carroll insisted on driving Coletti back to Bel Air. We all waved and laughed, as Carroll drove over the curb and blasted away! A new chapter with the legend had begun.

Back in Dearborn, Manfred Rumpel drafted up a list of roles and responsibilities for Advanced Powertrain,

THE LAST SHELBY COBRA

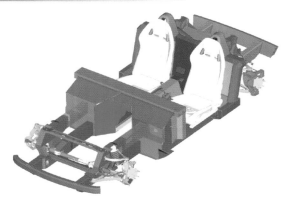

Daisy aluminum space frame.

Daisy powertrain: 6.4L V-10, torque tube and transaxle.

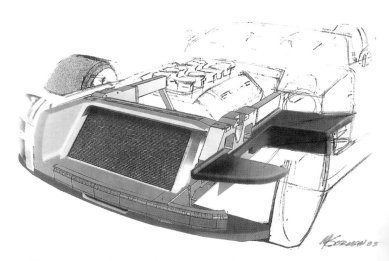

Sketch of V-10 velocity stacks. (Courtesy Bud Brutsman)

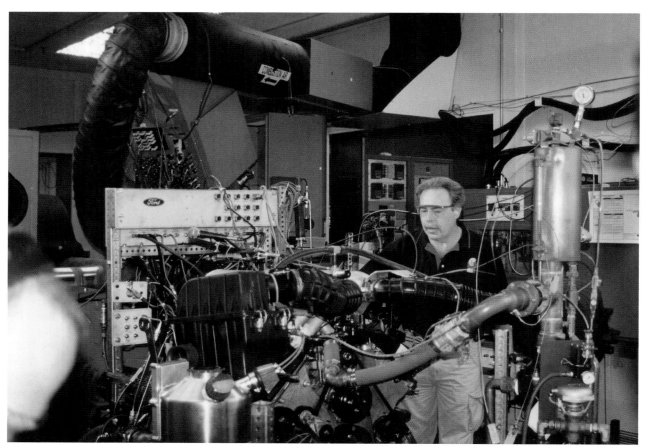

Greg Coleman dyno testing the V-10. (Courtesy Bud Brutsman)

the Valencia designers, and his engineering team, along with a timeline. Manfred began work on the chassis design and engineering. With the V-10 located behind the centerline of the front wheels and rear transaxle, Daisy would be a front mid-engine 'sport car' with 48/52 per cent front/rear weight distribution. The team designed a passenger compartment with more legroom than a Crown Victoria in a vehicle no longer than a Mazda Miata. In addition to the suspension, Daisy would use the Ford GT's transaxle, steering, brake, and cooling system components. The space-frame used many Ford GT aluminum castings and extrusions. All the design and analysis was done with CAD (computer aided design) and CAE (computer aided engineering). There was no time to build and test.

Shelby joking around in the Daisy seating buck with J Mays and Chris Theodore. (Courtesy Bud Brutsman)

Full size clay model of Daisy – note the engineering details in the background. (Courtesy Bud Brutsman)

Over at the Advanced Powertrain lab, Gregg Coleman and company were designing, sourcing, and assembling the V-10 for dynamometer testing. Without any good reason, other than it would look cool, I asked the team to develop an intake manifold with F1 racing inspired velocity stacks so they could be seen inside the hood scoop. The V-10 had to be calibrated and ready for installation by the end of July. Once designed and built, the 6.4L was put on test, producing a modest 605hp and 501lb/ft of torque!

Now that Hutting had a package, three-dimensional design could begin in the Valencia studio. As Hutting later stated, "The powertrain, the space frame and the suspension were all key elements in the design, although for the most part, you don't see them. These established our proportions and naturally led to a race-bred shape that evokes the original Shelby Cobra, without sharing a single dimension or proportion."

By the next trip to Valencia, the full size clay model was starting to take shape. Designing a modern Cobra was no easy challenge. Perhaps smarting a bit from some criticism regarding 'Retrofuturism,' J Mays and Hutting worked hard to create a minimalist, contemporary style. The front-end design proved

Theodore, Salenbauch, Mays, Shelby and Hutting discussing Daisy clay model. (Courtesy Bud Brutsman)

THE LAST SHELBY COBRA

especially difficult, as the signature round headlights and oval grill of the original were hard to translate into contemporary graphics. Key details, including the dominant grille opening, hood scoop, vertical bumper bars, and stacked lamps, front and rear, helped establish the historical connection to the original. Carroll was pleased with the progress, and the mean, aggressive stance that was taking form from the muscular fenders. J said "when you're setting out to tell a story about an automobile in a fresh, contemporary way, you're not actually looking to create beauty – you're looking to create meaning" – and Daisy looked mean!

Hutting stayed true to the package Manfred's team had created. As he said, "we started with the mechanicals – they set up the architecture for the body

Final test in Dearborn with Manfred and Chris.

Daisy chassis ready for shipment. (Courtesy Ford Motor Company)

CODENAME: DAISY

– and we went from there, interpreting the classic Cobras of the 1960s in a truly modern automobile." Richard only asked for one concession, which we quickly agreed to: He lowered the front chassis rails so that one could get a better view of the awesome V-10 with the hood tilted. Shelby was mesmerized by the CAD animation as the hood tilted to reveal the V-10. Hutting even incorporated aerodynamic suggestions, including front and side splitters, a rear diffuser, and, at my suggestion, F1 style 'barge boards' to manage air extraction from the engine compartment and wheelhouse.

We then moved to a plywood interior buck mounted on a steel surface plate. Cameras were rolling as Carroll climbed up and into the buck, quipping "I'm stiff in all the wrong places." That quote ended up on the cutting room floor, but kept everyone amused. At 6ft 2in, Shelby was amazed at the roominess – and no offset pedals to contend with. We all agreed to keep it simple: no paddle shifters, no air-conditioning or radio, and we even eliminated the exterior mirrors and door handles. Three hidden cameras would project a panoramic rear view in place of a mirror – a modern version of the prismatic mirrors used by racers in the '60s.

Shelby was joking and laughing the whole day, keeping the atmosphere light and the team engaged. Carroll had a magical way of inspiring people with his 'aw, shucks' Texas charm, and everyone took joy in pleasing him. We ended the day deciding that, in order to complete Daisy in time for the Detroit auto show, both the Valencia design team and Dearborn engineering team would have to work in parallel. To accomplish that, two space-frames would be required. One for shipment to Valencia where all the body and interior components could be fabricated and fitted. A second frame for assembly into a drivable chassis and testing in Dearborn. Team communication and coordination would be critical.

By August, the clay model had to be 'design frozen.' We had Carroll out for one last look. Mays led Shelby out to the courtyard and pulled the cover off the

Daisy before shipment to Carroll's shop for his first look. (Courtesy Bud Brutsman)

Completed Daisy at Shelby American shop in Gardena. (Courtesy Bud Brutsman)

49

One of the many shots of Carroll Shelby and Daisy taken at the photo shoot. (Courtesy Ford Motor Company)

final model, 'Dynoced' in silver, looking realistic, like a finished car. The look on Carroll's face said it all, "Oh, I'm really happy. Golly!" Now it was time to execute.

Metro Technologies Ltd fabricated the aluminum space frames, the same team that made the prototype Ford GT frames and fixtures. They milled Hutting's lowered front rails out of billet aluminum. The torsional stiffness of the frame fell a little short of its target: Corvette. I suggested that the bottom of the tunnel be fitted with a steel rather than aluminum plate. Sure enough, the stiffness increased from 8695 to 10,060lb/ft per degree, exceeding the Corvette's. The second frame went to Technosports Inc in Livonia, Michigan, where all the suspension, steering, brake, cooling, electrical, torque tube and transaxle components were fitted.

Next stop was the Advanced Powertrain Labs, for the V-10 installation. On my way back from a meeting in LA with Carroll, I stopped by to see the V-10 come to life. Things did not go quite as expected, as the car would not start, and then only sputtered to life after fiddling with the calibration. The team stayed the entire night trying to solve the problem. Finally, at 6.30 the next morning, they discovered two crossed wires, and Daisy fired right up. Later that day Greg Coleman got Carroll on the phone, started Daisy, and let him listen to the rumble.

"No runs, no drips, no errors," reported Greg.

"That's great guys," Shelby responded.

Back in Valencia, Hutting and the team were designing and fabricating every single body component for Daisy. Unlike most concept cars, built by outside fabricators, this would be done completely in-house. No ordinary push mobile, or 15mph electric cart, Daisy's body had to withstand race car stresses. Instead of a simple fiberglass shell, body panels were made of a double-wall, fiberglass foam core sandwich. The hood would be raised hydraulically, and, when closed, would be latched in place with electrically actuated deadbolts. The Sparco carbon fiber seats modified and hand trimmed. Body panels painted and striped. Bumpers, pillars and roll bars milled out of aluminum. Every component was designed and handmade, right down to the instrument cluster graphics and toggle switches.

Working on a tight schedule, the team completed Daisy just six days before Shelby's scheduled demo drives with select members of the press. Monday, Daisy arrived at Shelby's shop in Gardena, where J Mays met him to unveil the car.

"Oh dang," said Carroll. "That is really something. It just turned out beautiful didn't it?"

Tuesday was reserved for a photoshoot with Carroll and Daisy at Soundstage. The 'Trilogy' Ford

Shelby's first Cobra CSX2000 sold for $13.75M at auction in 2016. (Courtesy *Motor Trend*)

Shelby doing donuts in Daisy for the cameras. (Courtesy Bud Brutsman)

THE LAST SHELBY COBRA

Rides photo crew, filming Shelby at Irwindale Speedway. (Courtesy Bud Brutsman)

My turn with Daisy at Irwindale. (Courtesy Bud Brutsman)

Men in black after a day's filming with Daisy: Scott Strong, Richard Hutting, Carroll and Chris. (Courtesy Bud Brutsman)

GT and 2005 Mustang GT joined the photo shoot on Wednesday, and on Thursday the team got a chance to tune and debug the car. I tag-teamed with Mays and arrived at the Irwindale Speedway on Friday for the *Rides* video shoot, and press backgrounder on Saturday.

The video shoot at Irwindale Speedway, a banked half mile track, was a long and amazing affair. Bud Brutsman secured the track, and brought out the high definition video team, along with bodyguards to make sure our secret was safe. Carroll brought out the first Shelby Cobra CSX2000 (which sold for $13.75M at auction in 2016), and a polished aluminum bodied 427 Cobra. Shelby spent the morning driving both Cobras for the film crew, and then it was time for Daisy. Carroll did tracking shots on the oval and high speed runs around the short-banked track, but the best shots were of his smoky donuts in forward and reverse, captured by an overhead helicopter. Bud's film crew combined these with similar shots of the 427 so the viewer could watch Shelby magically transform the original Cobra into the new one through a cloud of smoke. I finished the day's driving so the crew could get more tracking shots of Daisy. It was then that I realized that even at 80 years of age, Shelby had retained his legendary driving skills. The slide throttles for the velocity stacks were sticking, making it hard to control speed, you were either at idle or wide open! Tracking shots at constant speed were almost impossible, and driving through the corners required two foot driving. Ol' Carroll might not see too well, but he could still handle a car! As the sun was setting, Carroll took the time to pose for photos with Daisy team members.

Before sunrise, Carroll and I were back at Irwindale for embargoed press previews. First up was Matt Stone, of *Motor Trend*, who'd been promised an exclusive article. I took Matt through the Daisy 'build book' to explain the technical details. Daisy was not the typical concept car. It was a running, high-performance, production feasibility study for the next Shelby Cobra. Shelby waited on the track below, to give Matt a ride in Daisy. Matt later wrote that it was of little concern to Shelby "that he was driving a multimillion-dollar hand-built prototype, as he stabs the gas and takes the racer's low line through a long, sweeping corner." We spent the rest of the day repeating the process, as other journalists arrived for their scheduled ride with Carroll and Daisy. By the time Shelby was done, Daisy had performed flawlessly for over 150 hard miles, ready for the trip to Detroit for its debut.

To make sure the entire press corps knew that something big was coming to the Detroit auto show, I suggested to our press flack, Dan Bedore, that we send

THE LAST SHELBY COBRA

out a daily email blast for the 12 days of Christmas, revealing a different component of Daisy each day. Dan went one better, and released the "Ten days before the Auto Show, the Blue Oval gave to me, a secret named 'Project Daisy'," concluding with: "Ten pistons a pounding, 'Nein!' Porsches passing, W-eight saving spaceframe, Seven spokes a wheelin', six forward speeds … fiiiiiiive hundred lb/ft, four massive Brembos, three clicks to 60, two Sparco seats, and a ride inspired by the Ford GT." Every automotive journalist got the message, although a few did complain about the email overload. The date for the Ford Press conference and big reveal was fast approaching. J Mays came up with the idea of the Ford 'Trilogy' where the Ford GT, and New Mustang would be painted in the same colors as Daisy: Tungsten Grey Metallic with Silver stripes. In fact, there was some precedence in the trilogy, as Ford had made a film in 1965 with Shelby, showing how the Mustang GT350, Cobra, and Ford GT were developed at Willow Springs Raceway. Still needing something for the introduction to the press conference, Bud Price asked me to explain why these cars were so significant. I told him the story of racing slot cars as a 13-year-old kid, and switching from Ferrari to the Ford GT when Ford won Le Mans.

Sunday, January 4, 2004, at the Cobo Arena in Detroit, the lights lowered at the press conference as the show opened with a video of young 'Bobby' watching a 1965 Mustang commercial on a black and white TV, followed by footage of Ford: 'Performance

Carroll Shelby and Bill Ford Jr at Ford's Detroit NAIAS Press Conference on January 4, 2004. (Courtesy Ford Motor Company)

CODENAME: DAISY

My favorite shot of Carroll and Daisy: sunrise on the banking of Irwindale. (Courtesy Bud Brutsman)

on Parade'; with images of GT350s, GT40s, Cobras, and Daytonas. Young Bobby is called up to dinner, after which he returns to the basement to play with his Corvette slot car on the orange shag carpeting. Mom yells, "Bedtime," and Bobby is off, but the Corvette slowly comes to life on the track with the noise of the crowd in the background. Suddenly, in the pits a Shelby GT350 fires up, then a Ford GT and finally a Cobra. As they all pass the Corvette, it flies off the track into a billboard. The closing scene has the Cobra slot car heading straight towards the audience. Fireworks go off at the sides of the stage as the Tungsten Mustang and Ford GT roll on stage. Finally, thunderous, huge fireworks erupt center stage, and Daisy launches onto the turntable with Carroll Shelby. The crowd roars with applause. Bill Ford Jr comes up to congratulate Shelby, and the press descends on them for a closer look. Shelby had said he couldn't wait for the press conference, "It's going to be the best day of my life." I'm sure it was, but there would be many more good days to come.

The reviews in the following days were outstanding. Carroll and Daisy were on the front page of the *Detroit News* and *Free Press*. There was even a cartoon in the comic section with a cherub pointing an

J Mays and Carroll Shelby with Daisy. (Courtesy Ford Motor Company)

CODENAME: DAISY

Dick Guindon cartoon of Daisy. (Courtesy *Detroit Free Press*)

arrow at a husband ogling Daisy, as his wife tries to fend off the cherub.

Carroll, J, and I spent Monday and Tuesday during press days conducting interviews, but Tuesday evening, we left for the basement of Bailey's Bar & Grill in Dearborn. Bud Brutsman had arranged for a simulcast of the premier of *Rides* on TLC. The entire team was there, including designers from Valencia, the Advanced Engineering and Powertrain teams, and suppliers from Metro Technologies and Technosports.

Daisy on the Dearborn handling track.

Carroll and Cleo sat on the couch watching the show, as we made fun of each other's cameos on the big screen. Shelby's flip phone rang. He looked at it, but didn't answer.

"Who is it?" I asked.

"Ford, I'm having too much fun to talk, I'll call back tomorrow." *Rides* became the top rated show on TLC, and Codename: Daisy the most watched episode on TLC for the next couple of years!

Friday night, before the NAIAS opens to the public, Detroit hosts a $400-a-head Charity Preview Ball, a black-tie affair, where Detroit's who's who drink champagne and stroll though the show to look at the new cars. Carroll sat at a small table at the Ford both for three hours, as Detroit celebrities lined up for autographed posters, press kits, or anything else they could get their hands on. Under the hot lights, I was worried about Carroll's health, and whispered to him that I thought he should take a break. "I learned a long time ago that when the gettin' is good, the good keep gettin'," he responded, as he continued to sign autographs and pose for photos under the sweltering lights. When the Ball came to an end, the Ford marketing people had forgotten to invite Shelby to dinner with the dealers. Seizing the opening, I invited Carroll to join us for dinner at the storied Detroit Athletic Club as a guest of ASC, where he continued to receive congratulations from all in attendance.

Carroll Shelby and Daisy made the covers of virtually every automotive magazine both in the US and abroad. *Autoweek* magazine awarded Daisy and J Mays 'Best in Show.' Matt Stone concluded his review by stating "There's one final reason Ford should – no, must – give us the Cobra: to put the final, iconic punctuation mark on Carroll Shelby's extraordinary life, with a car that's worthy of the name." Stoking the flames, Ford followed up with a 'Trilogy' commercial showing the new Mustang, Ford GT and Shelby Cobra racing through the streets of Detroit, with the city skyline in the background. Now all we had to do was get the program sold and into production.

Carroll and Chris at Bailey's.

Thomas L Bryant, Carroll and Cleo Shelby, and Chris Theodore. (Courtesy *Road & Track*)

Opposite March 2004 *Motor Trend* Cover. (Courtesy *Motor Trend*)

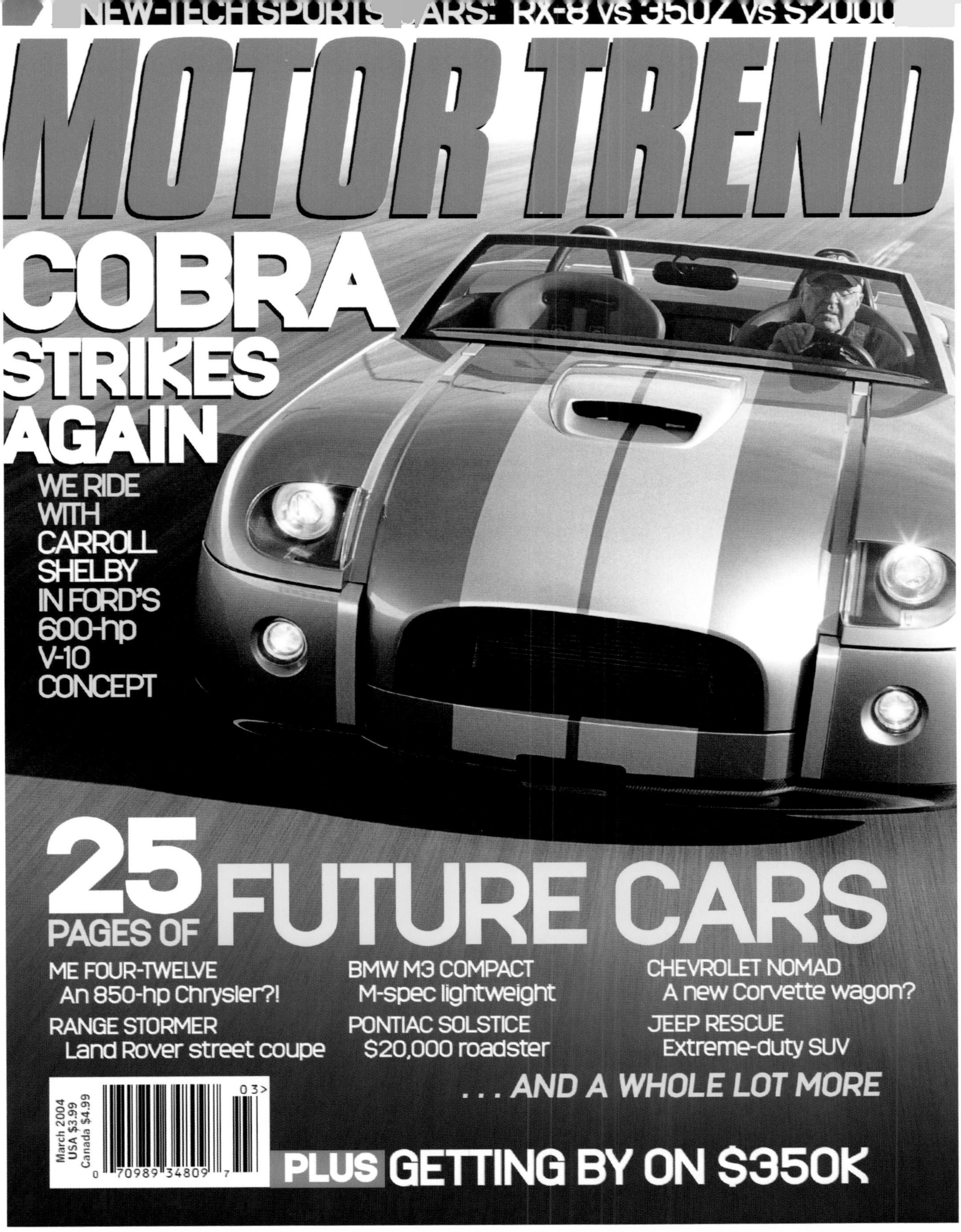
NEW-TECH SPORTS ARS: RX-8 vs 350Z vs S2000
MOTOR TREND
COBRA STRIKES AGAIN
WE RIDE WITH CARROLL SHELBY IN FORD'S 600-hp V-10 CONCEPT
25 PAGES OF FUTURE CARS
ME FOUR-TWELVE
An 850-hp Chrysler?!
RANGE STORMER
Land Rover street coupe
BMW M3 COMPACT
M-spec lightweight
PONTIAC SOLSTICE
$20,000 roadster
CHEVROLET NOMAD
A new Corvette wagon?
JEEP RESCUE
Extreme-duty SUV
. . . AND A WHOLE LOT MORE
March 2004
USA $3.99
Canada $4.99
03>
0 70989 34809 7
PLUS GETTING BY ON $350K

Chapter 6
SHELBY GR1

The New Year was off to a great start with the introduction of the Shelby Cobra concept and the success of the *Rides* show *Codename: Daisy*. We immediately began work on production feasibility and business case studies. Daisy had been moved to the Living Legends studio, where Camilo was assigned to address some of the critics' comments by making the design more emotional and sexy. The rest of the team tried to make the dream a production reality.

Back in 2003 when J and I decided to do the Cobra Concept, we also talked about following up with a coupe in the mold of the beautiful Shelby Daytona designed by Pete Brock. My first chance to tell Nick Scheele of our plans for 2005, came after I finished giving a speech on 'Changing at Ford GT Speed' at the CAR Management Briefing Seminar in Traverse City. Waiting for Scheele's speech to follow in the next session, I was doodling when Nick took a seat next to me. I passed my doodles along. Actually, these were doodles for a removable hardtop for Daisy, similar to some aerodynamic efforts on Cobras at Le Mans, before Pete Brock designed the Daytona. If you can't put an engine under glass, like the Ford GT, the next best thing is a beautiful spare wheel and tire.

The actual sketch for the 'Daytona' came about one day in November thanks to a young designer, George Saridakis. He had seen the Daisy rolling chassis at the Irvine studio, and thought, "What can I do with this?" In a stroke of genius, he sketched three different views of a gorgeous coupe. When J saw them, he told George, "Don't change a thing," and sent them along to me. I immediately agreed – it was exactly what we were looking for. Rarely in my career have I seen a sketch that immediately screamed, "This is it!"

The next step was to get Carroll's seal of approval. On one of our trips out to the Irvine studio, Hermann Salenbauch and I drove up to the Bel-Air Country Club and met with Shelby. We sat outside the Country Club Grill, overlooking the LA skyline, and watched the members tee off. Everyone stopped by to say hello to

My doodles for a Daisy Coupe.

Carroll, and asked when he would next be out on the course. I didn't expect Shelby to still be golfing at age 81, but I should have known better. Hermann pulled George's sketches from a folder and asked if he would approve our turning it into a Shelby concept car. Carroll loved the photos and was flattered that we asked. He then went on to tell the story of the difficulties they

SHELBY GR1

GR1 front view in racing livery. (Courtesy Ford Motor Company)

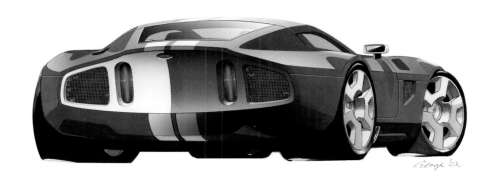

GR1 rear view. (Courtesy Ford Motor Company)

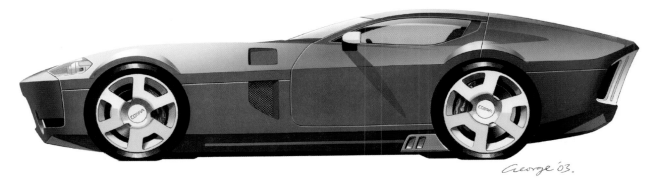

GR1 rendering by George Saridakis. (Courtesy Ford Motor Company)

had with Brock's design, having to add a huge spoiler to keep the rear wheels on the ground. Nevertheless, it was Brock's design that improved aerodynamics enough for the Cobra to compete at Le Mans, finishing 1st in class and 4th overall, with Dan Gurney at the wheel. In fact, Brock had difficulty convincing Phil Remington that a rear spoiler would be required. We told Carroll, that similar to the Ford GT, we would manage lift using under-car ground effects. Unlike Project Daisy, Carroll would not see the car again until August.

Back in Dearborn, Camilo was making Daisy more voluptuous. The British enthusiast magazine, CAR, loved Daisy, put it on the cover and declared it pure "Evil." Others thought it was too sinister, describing Daisy as "Pitbull meets Rottweiler." Walking into the studio, I noted that all work on Daisy had stopped. "What happened?" I inquired. "J is off Daisy, and on to the new blond!" someone replied, pointing to the coupe sketches. Still, I kept the finance team working on business plans for both cars.

With Carroll's approval, Manfred Rumpel and the Advanced Product Creation team were off and running again. The spare chassis from Daisy would be modified for the new coupe body style, adding a front crush structure for the longer overhang, roll bar for the roof, and revised fuel tank and fill system, among other detail changes. A new twist would be that a fiberglass rolling model would be unveiled at Pebble Beach on August 17.

Work on the full size clay model began in February at the Irvine Studio, under the direction of Henrik Fisker (yes, that Fisker). TLC/Discovery again documented the development – this time with emphasis on the design (ie, 'styling') process. George had the lead on creation of a model to match his sketches, but Henrik kept a close eye on the project to make sure the surfaces delivered what the sketches promised. In particular, the dramatic rear wheel arches had to be modeled perfectly. A 1mm change in the depth of the surface could move the reflections (Henrik called them 'ZED' lines) fore or aft by inches. Basically, the designers had to get the highlight centered above the rear wheelarch. This required a lot of finesse by the expert clay modelers at Irvine. To make matters even more difficult, somewhere along the line J and I thought it would be neat if the Daytona had a polished aluminum surface, like the 427 Cobra Carroll brought to Irwindale for the Project Daisy shoot. Any defects in the surface would distort reflections like a fun house mirror!

I do not recall why we didn't call the car the 'Shelby Daytona Concept' – probably because Chrysler still owned the trademark, or if not, perhaps it might have reminded enthusiasts of Shelby's Chrysler products. Nor do I recall who came up with the GR1 designation. Reportedly, the GR stood for 'Group Racing.' To me it didn't matter, so long as the name Shelby was attached to it.

Because we had an early start on GR1, and had the experience of Daisy behind us, along with the spare chassis, one would have expected the project to be less stressful. The challenge of a polished aluminum body, along with butterfly doors (which we should have used on the Ford GT) meant we had to get the running chassis completed by June 7, to allow time for creation of the aluminum body. There was no time to waste.

Development testing of the completed chassis was wrapping up by the end of June. It happened that it was North America's turn to host the Ford Global Product Development team, under the direction of Richard Parry-Jones, to share best practices. I volunteered to arrange the meeting for June 29 at the Dearborn Proving Grounds. Most of the international PD executives had never seen Daisy or GR1, let alone driven a Ford GT or the Cobra Concept. What better way to demonstrate the accomplishments of the small SVT and Advanced

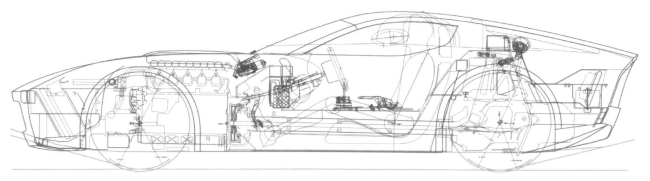

CAD drawing of Shelby GR1 on modified Daisy chassis.

SHELBY GR1

Clay model – front view ...

... and front ¾ view

THE LAST SHELBY COBRA

Rear ¾ view with 'Kamm' tail.

Rear overhead view; my favorite.

Opposite top GR1 technical briefing to Global Product Development team.

Opposite bottom Shelby GR1 chassis evaluation.

SHELBY GR1

Product Creation teams, than to present the technical details, and let them drive the completed GR1 chassis? We started the day with a technical briefing on the Ford GT, Cobra Concept, and the GR1 to come, then turned them loose on the test track.

The technical briefing explained the benefits of the small team concept, new technologies developed, and the ability to share a new set of supercar components. The track time, however, made for the most memorable gathering in the history of those meetings. The GR1 chassis drive was exhilarating – no pun intended – like driving a 600 horsepower go-kart!

Meanwhile, the fiberglass rolling model was coming along nicely at the Irvine studio. It was painted in the same 'Trilogy' colors that would become available on the 2006 Ford GT. Carroll was invited to Irvine for a

THE LAST SHELBY COBRA

first look at GR1 and a photo shoot for the press kit. Thursday night before the Pebble Beach Concours, Ford hosted a large press party at the Beach & Tennis Club, 'Ford's Celebration of Performance Art.' Sitting next to Daisy was the GR1 under a silk cover. J Mays invited everyone, including Carroll and Cleo, to gather round for the unveiling. With a flourish, J pulled off the cover announcing, "This is a gift to Carroll." The response to GR1 was universally positive. No controversy this time! George Saridakis was introduced as the designer. His parents, John (of Greek descent) and Brenda (Scottish), flew in from London to see what their son had created. Their faces beamed with pride.

Clockwise from top left
Cleo, Carroll, and J Mays. (Courtesy Andrew Gardner)

George Saridakis with his creation.
(Courtesy Andrew Gardner)

Shelby with the GR1 fiberglass model.
(Courtesy Ford Motor Company)

SHELBY GR1

Unlike Daisy, the body and interior of GR1 would be built by an outside fabrication house, Aria, in Irvine, where the completed chassis was sent. I wanted to show off the superplastic aluminum forming technology used on the Ford GT. Body panels were formed by Superform in Riverside. In a moment of irrational exuberance, I suggested that we repeat the trilogy for NAIAS with Ford GT, Daisy and GR1 all in polished aluminum. Fortunately, cooler heads and the budget prevailed. While the body panels were indeed superplastic formed (a process where aluminum sheet is heated to 500 degrees centigrade and then slowly blow formed over the tool with hot air), the front fascia exhibited too much spring-back. No amount of heating or beating could persuade the aluminum to take the proper form. Consequently, the entire front had to be NC milled out of a block of aluminum – an expensive process.

While fabrication continued, the business cases were coming together. Three alternatives were first considered: Cobra roadster ($150M), Ford GT Mk II continuation and Cobra ($180M), and GR1 ($160M). Assumptions for the roadster and coupe were a $99,000 MSRP and 1500 units a year for 4 years. The roadster would cost about $68,000 to build, while the coupe would cost nearly $72,000. Neither would break even; the roadster losing $8M and coupe losing $14M over the proposed four-year life of the program. With the addition of a 600hp Mk II Ford GT continuation priced at $170,000, however, nearly a $12M profit could be realized.

We also studied a supercharged V-8 vs a naturally-aspirated V-10. Investment for a 600 horsepower V-8 would, of course, be lower than the cost of a new V-10, but the V-10 would cost about $3000 less to manufacture. The only scenario that made financial sense was to continue with the Ford GT Mk II (at 500 units a year) along with either the roadster or coupe. Marketing, however, did not want to continue the Ford GT, as they had promised to limit total Ford GT production to 4500 units, nor did they want to offer two high-performance coupes at the same.

We studied one final scenario in late September: Launch GR1 while the Ford GT production was ending with $119,000 MSRP and build 1500 units per year; follow with a Ford GT Mk II at $170,000 MSRP building 500 per year; and finally introduce the roadster at a $99,000 MSRP for another 500 units per year. This

J Mays introducing the Shelby GR1 at the 2005 NAIAS. (Courtesy Ford Motor Company)

would have generated $160M in cumulative cost contribution, essentially breaking even. No production program decisions were made in 2004. Perhaps if a higher price had been assumed we could have made a better business case, but sales volumes would have been uncertain. An historic product opportunity was lost. In retrospect, the 2008 'Great Recession' would likely have killed these cars off anyway.

As the end of 2004 approached, John Coletti, Mike Zevalkink, and I announced our plans to retire. Coletti put it this way: "Once you've done the Ford GT, there is no higher mountain to climb!" The Shelby GR1 was completed on schedule for the 2005 NAIAS auto show and I came back to see its press introduction. At the finale of the Ford press conference, huge flames shot upward as the polished GR1 rolled out into the surrealistic blue spotlights as streamers and confetti floated down from the rafters. Later, walking back to the Ford display stand, I saw Hermann and Carroll, and went over to congratulate them. Hermann had been thoughtful enough to have a model of the GR1 made for me, which Shelby was signing as I approached, "Your friend for many years & many more." It was a special way to remember my time with both Carroll and Ford.

Before I retired in 2004, there was one last thing I needed to accomplish: Getting Shelby's name on a Ford Mustang!

GR1 gift from Shelby and Salenbauch.

Press photo of Shelby GR1 in the California desert. (Courtesy Ford Motor Company)

Chapter 7
PROJECT CONDOR

The story of reuniting Mustang and Shelby began in 1999 with plans to replace the old Mustang, whose 'Fox' platform dated back to the 1978 Ford Fairmont. That this Mustang had survived for so long was a testament to the strength of the brand, as it would be the last 'horse' standing in the pony car market. To do a new 2004 world class Mustang would require a leap of faith, and investment in a new rear-wheel drive (RWD) platform. Ford had spent billions developing the 'DEW98' RWD platform for the Lincoln LS, Jaguar S-type, and Thunderbird, only to cancel larger, more expensive derivative models. 'DEW98' was a good foundation, but too expensive for a Mustang that needed a base price starting below $20,000. The Capstone plan we developed called for a new platform, dubbed 'D2c,' that would start with the DEW floorpan, but use a MacPherson strut front suspension and a new, low-cost independent 'twist blade' rear suspension, similar to that used in Europe on the AWD Mondeo wagon and Jaguar X-type. The new RWD platform would be designed to be scalable for a host of new products, including a Ford sport sedan, entry-level Lincoln and even the Australian Falcon.

To make sure everyone was on the same page, I had the team write future road tests of the new Mustang and other products envisioned. Providing a little 'guidance,' I wanted Mustang to be light, lean, and have the dynamic capabilities of the 3 Series BMW. Aesthetics were to be a modern interpretation of the original 1965 Mustang Fastback, with long hood, short deck, lean body sides and strong wheel flares. Ideally, the fastback would conceal a hatch for increased utility, and there would be a convertible model as well. The imaginary 'Road & Test' review became a guiding light for the team, and Doug Gaffka's designers came up with some contemporary ideas that captured the Mustang's essence. In addition, a 'Galaxy 300' sports sedan would also challenge the BMW 3 Series, which according to its 'road test' was an excellent value proposition with world-class handling capabilities.

Hermann Salenbauch and Hau Thai-Tang were early leaders of the team, heading up design and development. Hermann, a gregarious Bavarian, had come to us from BMW via Rover, while Hau, a Vietnamese immigrant (one of the last children out of Saigon during the war) was one of Ford's brightest young engineers. He had interned on Bobby Rahal's Indy car team, and worked on the Lincoln LS. Only now, as I write this do I realize that we had entrusted an American icon to immigrant engineers! I have noticed over the years that sometimes an outsider can better capture the essence of a culture than those who live in it. This seemed to be true of designers as well. In any event, as work continued, the S197 was coming along nicely. Financial realities forced some cuts, but the team was able to hang on to a strong platform foundation. The hatchback had to go for reasons of cost, weight, and body torsional rigidity, and we had to make some compromises on powertrain. The planned 3.5L four-valve V-6 would have to wait, but the crude 3.8L 90-degree V-6 was replaced by the 4.0L 60-degree 'Cologne' V-6 producing 202hp. Dreams of utilizing aluminum spray bore technology, developed by Ford Research, to create a short stroke aluminum block V-8 would have to wait. Instead, the 4.6L V-8 was fitted with SOHC three-valve heads and produced a respectable 300hp. For the new Mustang to have a strong stance, wheels and tires needed to fill the wheel openings and be as flush as possible. Making sure this happened required close co-operation between design and engineering, as the old Ford design standards were unacceptable. We scheduled 'sunrise meetings' at 6AM in the studio every day, to hammer out all the design details. The design theme was coming together: bold, aggressive, but with modern interpretations of design cues from the best Mustangs of years past.

By mid-2002, the Mustang design was pretty well locked in, and J Mays and I thought it would be a good idea to give Mustang enthusiasts an idea of what was coming at the Detroit Auto Show. It would also be a

THE LAST SHELBY COBRA

Early Mustang concept sketch.

'Galaxy 300' sports sedan sketch.

good way to get feedback from loyal owners. Market research had shown that we had a winner on our hands with 'sporty car intenders,' but nothing is more important than your loyal customer base. I suggested a short wheelbase version of the production design theme, with a sectioned body, 1963 Mustang II show car front end, supercharged V-8 and Shelby GT500 graphics (fortunately, my sketch is long gone). We were still getting passive resistance regarding use of the Shelby name, so that would have to wait, but we did decide to do a SWB convertible and LWB coupe for Detroit. One day when J and I were walking though the studios, we noticed a fantastic rendering of a Mustang GT-R with a hexagonal grill reminiscent of the 1968 Shelby GT500.

"Why didn't we think of that for the production car?" I asked rhetorically.

"Too late, we'll save it for the Shelby version," replied J.

Mays assigned the concept car design to Richard Hutting and the Valencia studio. You can see hints of the 1968 Shelby GT500 front end in the concept cars.

Building the concept cars turned out to be a good exercise. What had looked good in clay did not transfer well when we saw the concepts in the flesh (particularly the convertible). The surfaces on the body sides and rear deck were too flat, causing the cars to look like folded cutouts. The solid red paint of the convertible accentuated the condition, as solid color paints do not 'flop' like metallics, lessening the apparent surface highlights. Hutting's team spent many long hours before the concept reveal to make the surfaces acceptable.

Mustang Concept sketch with GT500 graphics. (Courtesy Ford Motor Company)

PROJECT CONDOR

Sketch proposal for Mustang Concept Coupe. (Courtesy Ford Motor Company)

We now knew that the design staff in Dearborn, led by Larry Erickson, would have to go back and make the surfaces a little more 'full' on the production cars. Nevertheless, both concepts were hits at the show.

Later that year, I moved to head Advanced Product Creation (APC), and began developing a Lincoln revival plan while also developing new Ford Performance and SVT products. I took Hermann with me to head up APC, leaving Hau as Chief Engineer of the S197. The new head of Product Development managed to convince Nick Scheele and the CFO that he could save Ford $100 per vehicle by eliminating the independent rear suspension. With less than a year and a half until production, they initiated a crash program to accommodate a three-link live-axle rear suspension. Ironically, I later learned that the live axle ended up costing nearly $100 more than the planned independent rear suspension.

Now Coletti had introduced the first independent rear suspension on the 1999 SVT Mustang, and the outgoing 'Terminator' SVT Mustang was powered by a 390hp supercharged 4.6L V-8. Obviously, the new car, code-named 'Condor,' would have to outperform its predecessor in all aspects. Increasing power would be the relatively simple task of making a more affordable version of the Ford GT motor for the Mustang and

Convertible Mustang GT Concept. (Courtesy Ford Motor Company)

Mustang GT Coupe Concept. (Courtesy Ford Motor Company)

Got my wish: spare under glass! (Courtesy Ford Motor Company)

Shelby testing a GT500 prototype. (Courtesy Ford Motor Company)

soon-to-be stillborn Lightning (codename 'Tomcat'). A wet-sump, cast iron 5.4L block would replace the Ford GT's dry-sump aluminum block, and the expensive Lysholm supercharger replaced with an Eaton Roots-type blower. Output would be more than adequate, with 500 horsepower and 480lb/ft of torque on tap. The bigger challenge would be to put an independent rear suspension (IRS) in place of the live axle.

We put Manfred Rumpel and his team on the task of helping SVT come up with an IRS that could handle the high torque, and be assembled on-line at Mustang's Flat Rock assembly plant. The team came up with a novel 'internal link' design that avoided infringing on a BMW patent, and utilized the same spring seats, shock mounting locations, and fuel tank as the base Mustang. The only change to the Mustang body would be a weld-on bracket to the rear frame rails for attachment of the suspension module. SVT built two prototypes and put them on test with encouraging results. One went out west, and Shelby spent over five hours with the car, delighted with the performance. Carroll succinctly put it: "Wow!"

Of course, they left the yellow prototype behind, and it now resides at Shelby American in Las Vegas.

Two things that remained unsettled were the exterior design and name for Condor. It seemed obvious to me that Condor was a Shelby GT500. One might have assumed a potential conflict between an SVT developed product and use of the Shelby name, but nothing could be further from the truth. Coletti had long embraced the idea of bringing Shelby back into the fold, as did the SVT team and the design team. The resistance seemed to be coming from sales and marketing, along with the new head of Product Development. To flush out the resistance, on December 1, 2003, I sent a confidential letter to all interested parties and copied the president, suggesting a potential licensing agreement for Shelby's trademark names. This agreement would not obligate Ford to use them, but would secure their availability for future use. To my amazement, I received a note back from sales and marketing that the Shelby name wasn't worth more than $200. They preferred we develop a Bullitt model instead (these were the same people who wouldn't approve the 2001 model year Bullitt Mustang unless we offered it in black and blue, in addition to the correct Dark Highland Green!). Worse yet, the head of Product Development added a note in the corner of the letter to sales and marketing that said, "We need to move to the next generation and thank Carroll for all of his help with the prior one."

Not one to let things go, I followed up by sending an article that caught my attention in a business journal. It featured a photo of Pamela Anderson in a halter top and cut-offs, and cited CNW market research that Carroll Shelby tied with Pamela Anderson for first place with new vehicle intenders who would be much more likely to buy a car endorsed by them! Digging further into the data, Art Spinella, president of CNW Marketing Research, said, "Shelby is one of the few names in the automotive business that's recognized in some capacity by nearly every age group." Shelby's name beat out Petty, Harley-Davidson, Chapman, Earnhardt, Calvin Klein, Ralph Lauren, Mario Andretti and Lee Iacocca.

I told Carroll I was having trouble getting buy-in for the Shelby GT500, but would continue pushing. The resistance continued. It was not long after, however, that in staff meetings, The President of Ford North America would ask how negotiations were going regarding a Shelby license. "We're working on it," would be the response from the head of Product Development, hoping the issue would be forgotten. I checked with Carroll, no-one had approached him about licensing. Weeks later, the President asked again. "Still working on it," was the reply. The third time the President asked and received the same response. This time he became visibly agitated and ordered a meeting with Carroll to "get the deal done!"

This order was quite uncharacteristic of the president, as he did not usually get involved in product or marketing decisions. Carroll met with the EVP in Las Vegas. "How did it go?" I asked. "He wouldn't even look me in the eye. I didn't even bother to negotiate; I just took the deal and walked away." Only years later did Carroll tell me that, in exasperation, he had called Edsel and asked, "Why won't they let me put my name on a car?" Edsel must have said something. Had he not, there never would have been a 2007 Shelby GT500. Thank you Edsel!

With the name settled, Doug Gaffka's team started on making Condor into a Shelby GT500. The hood was raised 1in and functional vents added. The shape of the grille and front fascia inspired by the 1968 Shelby GT500, and the new front splitter derived from the GT-R show car. In back, they added a functional spoiler along with a new rear fascia that incorporated a diffuser with cutouts for the 3in stainless exhaust pipes. Blacked out ground effects gave the Shelby a longer, lower appearance, along with Ford GT inspired 19in cast aluminum wheels. New cobra snake badging adorned front, side and rear, along with 'skunk stripes' down the center – so no-one would mistake this Shelby for anything else. The interior received similar upgrades.

Before I retired, there was one more battle to fight, the independent rear suspension. By October of 2004, testing of the Shelby had proven quite successful.

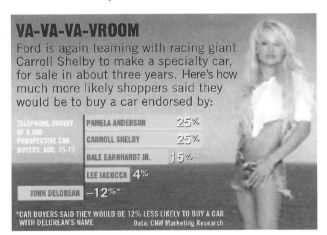

Shelby & Pamela Anderson tie for endorsement popularity!
(Courtesy Ray Vella)

THE LAST SHELBY COBRA

The increased performance provided by the GT500 did increase stress on the Mustang body, and a few areas were identified that could be strengthened with the minor improvements to the front and rear structure. Despite high confidence, the EVP decided that the IRS was "too risky" and ordered the SVT team to study a live rear axle alternative. On November 10, 2004, the team presented the projected effects of a live rear axle on performance. Despite a degradation in acceleration due to wheel hop and slightly poorer lateral acceleration, the EVP refused to sign long-lead funding for the IRS, and directed the team to change to a live rear axle and "improve" their performance targets.

The battle had been lost, but the war had been won. After seven years of trying, Shelby's name would appear on a production Ford.

Ford introduced the Shelby GT500 at the 2005 New York Auto Show, along with previews to the automotive press. The car received strong reviews and made the covers of *Motor Trend* and others with titles like "Shelby GT500! Ford's Bad-Boy is back … and meaner than ever." Yes, journalists noted the absence of IRS, but the Shelby name, performance and bang-for-the-buck could not be beat at $40,000. It would take another seven years before Mustang was fitted with IRS. Hau and Carroll toed the party line that drag racers preferred live axles for durability and ease of modification. Carroll had learned long before not to get involved with company politics. As he said, "I'm tickled and delighted to see the top-of-the-line Mustang be a Shelby."

If there were any lingering doubts about the power of the Shelby name, it was dispelled at Barrett-Jackson in Scottsdale the following January, when the honor of purchasing the first Shelby GT500 was

Carroll at the Shelby GT500 NY introduction. (Courtesy Ford Motor Company)

PROJECT CONDOR

auctioned off for $600,000 with all proceeds going to the Carroll Shelby Foundation. Deliveries began in May of 2006, with such strong demand that most sold at a substantial premium.

May 2005 *Motor Trend* Cover. (Courtesy *Motor Trend*)

Chapter 8
SHELBY GTS

You usually learn the difference between business relationships and true friendships after retirement. Business relationships fade, but friendships are forever. It was not long after I retired that Carroll gave me a call, and asked for my home address.

"Be glad to, but why do you need it?"

"Aw, I usually send fruit cakes out for Christmas, but my order got messed up and I need your home address to send something new," replied Shelby. Not recognizing the set up, I received a card the following week.

"Your Christmas Ham is now being made to order."

Of course, when I opened it, I saw the following:

The punchline!

That was Carroll! I called to wish him a Merry Christmas, and we shared a good laugh. The continuation of our friendship only grew stronger over the years.

Early in 2005, Shelby gave me a call, "Chris, would you be interested in running Shelby American in Las Vegas for me?" I was flabbergasted, and fumbled around for words.

"Carroll, that would be wonderful, but I'm not sure I can do it. Let me think about it and I'll get back to you." This was an opportunity I didn't want to miss, but with two young boys in school and both sets of elderly parents in Michigan, we just couldn't relocate the family. Carroll and I talked about commuting between Detroit,

LA, and LV. In the end I concluded I couldn't do justice to either Shelby American or my family. I still think about all the things we might have accomplished and the good times we would have had. Carroll and I agreed to stay close, as surely we would be able to do something together in the future. Years later, I stopped in to see Mike Zevalkink, and he mentioned his one regret: he too turned the job down, and for similar reasons.

Later that year, I sensed that my friends at ASC would be running into stiff headwinds, and thought I might be able to help. ASC (originally American Sunroof Corporation, and since renamed American Specialty Cars) had been founded by Heinz Prechter, a charismatic German immigrant. In true rags-to-riches fashion, Heinz came to America with ambition, a toolbox and $11 in his pocket. He started installing sunroofs in a two-car garage next to George Barris in Los Angeles. A larger than life personality, and true entrepreneur, Heinz sold his first OE program to Ford, installing 'moonroofs' on Lincolns. The business thrived, and Heinz became a captain of industry and 'kingmaker,' literally launching George HW Bush's presidential campaign from the diving board of his home swimming pool – a former Fisher Brother's mansion on Grosse Isle (yes, the founders of GM's Fisher Body). Unbeknownst to all but his family and closest friends, Heinz had hidden his manic depression behind a façade of boundless energy and optimism. Sadly, he took his own life in 2001.

I had met Heinz nearly 20 years earlier, working as Vice President of Engineering at Heinz' chief competitor, Cars & Concepts. To me, I thought of Heinz as the enemy. Where we struggled to sell programs to the OEs with sketches and proposals, Heinz beat us out of many contracts through his personal contacts with industry leaders and clever ways of breaking through the bureaucracy. For example, parking a concept car in the driveway of the Ford company president one night with the lights left on! We won our share of programs, including the 1983 Mustang Convertible, but it always seemed that Heinz was one step ahead of us. As I came to know Heinz, however, I learned what a genuine and warm individual he was. When we weren't competing, we broke bread at his magnificent home, or went for a ride on his yacht. When I was at Chrysler, Heinz introduced me to my counterparts at German OEs and suppliers, even inviting me to dinner with what became the new management team at BMW. Later, Heinz acted as matchmaker, for better or worse, between Daimler-Benz and both Ford and Chrysler. Fortunately, Ford recognized Daimler's overture as a takeover, while at Chrysler, CEO Bob Eaton fell for the 'merger of equals' pitch.

With the tragic passing of Heinz, ASC was sold to Questor, a private equity company, and Paul Wilbur, a former Chrysler colleague, had taken the job as CEO. Initially ASC prospered and landed a huge job to manufacture the Chevrolet SSR, and two more GM specialty car programs. The SSR was a retro hot rod pick-up with a retractable hardtop and hard tonneau, making it more of a show car than practical pick-up. It was heavy and underpowered the first year, and I suspected that its days were numbered. GM was heading into financial difficulties, and I feared that the SSR might be canceled. Having feelings for Heinz' legacy, as well as friends at ASC, I hired on as Vice Chairman that October, in hope of garnering new business. Sure enough, GM announced a couple of weeks later the closing of three plants, including the Lansing plant where the SSR was built. ASC revenues fell 90% within the year, and new business became the priority.

To my pleasant surprise, on October 19, 2005, I received a congratulatory note from Bill Ford stating,: "We'll be anxious to see the touch of American Specialty Cars on the continuing evolution of some great American cars, like the Ford Mustang." I immediately responded with a note thanking Bill and requesting a summit meeting to explore areas of mutual interest. A lot had changed at Ford in the ten months since my retirement. Ford, like GM, was going through a restructuring, and Mark Fields had been appointed President of the Americas. Fortunately, the old head of Product Development quit Ford for 'greener pastures', and Derrick Kuzak was named as his replacement (I had recommended Derrick to Bill Ford when I retired – not that I had any influence in the matter), and Hau Thai-Tang had been promoted to Director of Advanced Product Creation.

The summit meeting was scheduled for November 21. As expected, the assignment was given to Derrick and Hau for initial discussions. The purpose was to "explore potential areas of mutual interest." The Ford restructuring plan included the closing of the Wixom plant and ending production of the Ford GT, Lincoln LS, and Thunderbird. The Ford GT was high on my list of opportunities, so I called Carroll.

"Would you be interested in partnering with ASC on a Shelby continuation of the Ford GT?"

"Hell yes!" was the response.

"Great," I answered. "I'll include a proposal for a low-volume Shelby GT, and we'll see what they say."

In the meantime, Carroll sent a letter to Mark Fields saying that "with the scheduled build-out of the Ford GT next year, I'd like to propose some follow-on limited production Shelby/Ford Products … I look forward to discussing this with you the next time we get together."

At the meeting, we opened with three restructuring opportunities: 1. Ford GT continuation at ASC/Shelby, 2. Thunderbird and Lincolns LS derivatives produced by ASC for Ford, and 3. continuation of other models that were being phased out (eg Ranger, Sport Trac, and Crown Victoria). We also discussed other 'buzz' product actions, and formation of a strategic partnership for the development of new concepts. To my delight, Derrick expressed strong interest in the Shelby GT, so long as the badge said "Powered by Ford" – a historic nod to Shelbys of the past. A few other specialty projects were also selected for further study.

That evening, I called Carroll to give him the download from the summit. He was thrilled. That same day he received a call from Mark Fields' administrative assistant stating that Mark would get together with Carroll in January, "once he got his ducks in line." Carroll also mentioned that Ford Marketing was talking to him about a number of other projects, including a Shelby Expedition. Shelby said he would send his "marketing and licensing" guy to follow up with us.

My head was spinning as I started to put together a more detailed proposal. First out would be a Shelby GT Mk II with modifications inspired by the 1996 Le Mans winning cars, but I was already dreaming of other derivates further down the road. ASC and Shelby would purchase the rights to the Ford GT technology, tooling, and equipment. At the end of the 2006 Ford GT model run, the higher priced ($250K), 'low-volume' (250 units/year) Shelby Mark II would be distributed through participating Ford dealers. As a low-volume manufacturer, Shelby/ASC would be responsible for development, certification to the new safety and emission standards, and manufacture at an ASC facility. Power would be upgraded to a minimum of 600 horsepower, suspension revised, with new wheels and tires fitted to address the weakest link of the Ford GT (the Goodyear F1 tires). I had one of our in-house designers, Doug Ungemach, render the Shelby GT I envisioned, and sent the proposal to Carroll for review before the meeting.

We met with Carroll's marketing guy, none other than Joe Jacuzzi (of the Jacuzzi whirlpool family, and currently serving as Executive Director, Vehicle Brand Communications at GM) on December 21. Joe loved the proposal, and also shared with us the proposal he was working on for Mustang. Amy Boylan and Gary Patterson had met with Ford on September 15, 2005 to pitch some Shelby product ideas, including Shelby Mustangs and a performance truck. Ford didn't bite on their proposals, but mentioned that Hertz had approached them about re-creating the original GT350H. Because Ford didn't have the rights to the GT350H trademark, they turned Hertz down. Following up on that lead, Shelby American presented a CS6 supercharged V-6 Mustang to Hertz. They liked the proposal, but wanted a V-8

Shelby Mk II design proposal.

SHELBY GTS

Rear view of proposed Shelby Mk II.

and the Shelby hood, but not the front fascia because it was too low for their transporters. Shelby American cut an exclusive deal with Hertz for gold-on-black painted Shelby GT-H models commemorating the 40th anniversary of the original performance 'rental car.' It would be fitted with Ford Racing performance parts that added 19hp to the V-8 and improved handling, and it would use the California Special front fascia. Joe and I laughed about the deal, since we knew that once Ford dealers saw the rental cars at the 2006 New York Auto Show, they'd want a version of their own to sell.

I sent Carroll a note the next day regarding the outcome of our meeting. Carroll asked that I take the lead on the Shelby GT with Ford, so he could focus on some of the other deals that were in the works. We agreed that Carroll would take the proposal to Fields in January at the Detroit Auto Show.

Carroll obviously had a good meeting with Mark, because on January 26 I received a note from Hau, requesting a detailed proposal by February 10 to include: business framework, tooling purchase, required technical support, powertrain purchase,

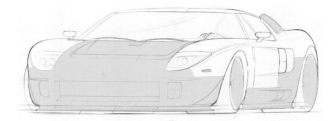
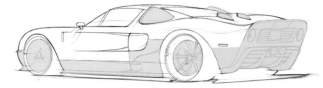

Proposed design changes for Mk II.

1966 and 2006 Hertz Shelby GT-H. (Courtesy Shelby American)

service parts, and timing. Fred Goodnow and Mike Isaacs would address product and business questions, respectively. Delighted, we put together a thorough business and technical presentation. We made our 24-slide presentation to Derrick Kuzak, Hau, Fred, Mike, and representatives from Legal, Finance, Manufacturing, and Sales and Marketing. Our goal was to agree to a Memorandum of Understanding by April 1, with purchase completion by October 1. At the end of the meeting, Derrick, and virtually all the other attendees, seemed quite interested. Derrick asked for more detail regarding other Ford GT derivatives and a product cycle plan. He also suggested adding a Targa model. The lone representative from sales and marketing, however, objected: "We promised Ford GT owners that only 4500 units would be built." Arguments pointing out that the higher price, Shelby name and lower volume would only enhance brand value fell on deaf ears. Derrick pressed on, "Why don't you survey Ford GT owners, and see what they think?" It was agreed that we would continue the study and Ford would provide all the information necessary to complete a MOU, and that we would get back together once the survey of owners had been completed.

Both the Ford and ASC teams worked hard to put a comprehensive plan together, but worried about the apparent intransigence of Sales and Marketing. To help work around this potential roadblock, Paul Wilbur sent a note to Mark Fields requesting a meeting to explain the strategic benefits of such an agreement to Ford: a 'no cost' way of keeping Ford halo products on the market. To the best of my knowledge, Fields never accepted a meeting, apparently choosing to let the team sort it out. The showdown meeting was scheduled for March 29.

Shelby/ASC presented a 53-page plan with story boards that covered every detail: product positioning and differentiation, sales and distribution, service parts and order fulfilment (including current Ford GT parts), warranty and business framework – including the formation and capitalization of a new legal entity, tentatively called 'Carroll Shelby Cars.' Most interesting, of course, was the product and cycle plan, with the Shelby GT Mk II launched in February of 2007, followed by the Targa in August of 2009, and Mark IV in July of 2011.

I was most excited about the Shelby GT Mk IV. Goodnow and I had long discussed the benefits of moving the radiators to the rear. One day on the way to ASC, with no front radiator to contend with, I envisioned the Shelby GT with F1 inspired adjustable front wings that encircled the obligatory bumper beam. It also finally occurred to me how to solve the wrap-over door issue with a diagonal butterfly hinge mechanism. Finally, twin-turbos would lower the center of gravity by

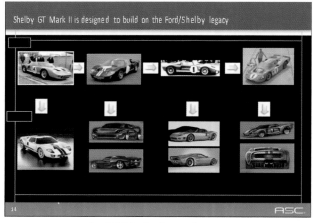

Ford GT/Shelby legacy line-up.

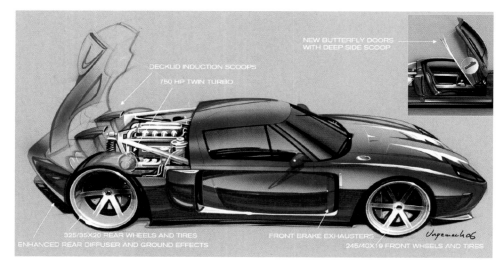

Shelby GT Mk IV with 750hp twin-turbo and butterfly doors.

Shelby GT Mk IV with integral front wing and rear radiators.

nearly an inch for better handling, while eliminating the 100hp parasitic loss of the supercharger, so that 750 horsepower could be reliably achieved.

Finally, we pointed out the potential for future opportunities with updated versions of the Shelby Cobra and Daytona, similar to the Daisy and GR1 concepts of 2004 and 2005. Derrick and the rest of the team were drooling over the future product renderings. All but the head of Sales and Marketing. He adamantly objected to the plan, and cited his survey of Ford GT owners.

"How many did you survey?" inquired Derrick.

"Nine, and two objected to a new Ford GT." Derrick rolled his eyes, but knew that politically the issue was dead.

Disappointed, but never wanting to let go, I had an opportunity to check the integrity of the survey at the Ford GT National Rally that August. Approximately 70 Ford GTs and 200 owners converged in Dearborn for the three-day event. Fred, Camilo, and Jamal were invited to speak at Ford's Conference and Events Center. Fred asked Neil Ressler and me to come up and say a few words. Prepared, I got up and presented the Shelby GT renderings, and then asked for a show of hands from those who would be interested. Everyone in sight raised their hand. I then specifically encouraged anyone who disagreed to raise their hand. There were, in fact, two dissenters. I had the chance to follow up with one of the dissenters, who owned seven Ford GTs (one in each color, plus a Heritage edition, and one to drive). He viewed the six hermetically sealed cars as investments! I asked him if when surveyed, he had been informed of the higher price and limited volume, and if it would have made any difference in his opinion. He hadn't been told, and said he would have answered differently. The number two man in marketing saw and heard all of this, but I knew his boss would never learn of it. A decade later, the Ford GT would return on its 50th anniversary.

Fortunately, things were going better for the Mustang-based Shelby GT. Five hundred 2006 GT-H automatic transmission coupes were produced at Shelby American (somehow four manuals snuck into the mix), and put into premium rental service, then another 500 Shelby GT-H convertibles were added to the fleet. Dealers began calling Ford for their own Shelby GT to sell. Bowing to dealer pressure, the Ford Shelby GT was rushed into production. The Ford Shelby GT was essentially the same as the Hertz cars, with Ford Racing cold air induction, suspension package, exhaust, 3.55:1 rear axle, and the five-speed manual became an available option. Appearance changes included deletion of the rear spoiler, a retro GT350 inspired hood scoop, 'skunk' and rocker stripes along with Shelby badging, inside and out. Press reviews were enthusiastic. *Motor Trend* noted that "the car carries legit Shelby cachet and is a surefire collectible … there's tangible content and performance improvement over a standard GT, at a price that makes sense." Like the Shelby GT500 the year before, this one raised $600,000 at Barrett-Jackson for the Carroll Shelby Foundation. In all, 5657 Shelby GTs were built at Shelby American in 2007 and another 2214 in 2008. Combined with the 8583 Shelby GT500s built at Ford's Mustang Flat Rock plant, times were good and Shelby was on a roll!

2007 Ford Shelby GTH offered in black and white, with silver stripes. (Courtesy Ford Motor Company)

Chapter 9
SUPER SNAKES

The year 2007 was shaping up to be a great one for Shelby, and an interesting one for me, as well. Hertz GT-H and Ford Shelby GTs were keeping the Shelby American facility humming, royalties were coming in from Shelby GT500 sales, new Shelby cars were in the works, and it would all start in Scottsdale, Arizona with perhaps the most significant Barrett-Jackson auction of all time, scheduled for January 13-21. I had been a regular attendee of the Scottsdale auction, ever since Chrysler became an early sponsor. Barrett-Jackson was a great venue to see fabulous cars, see old friends, and meet with automotive legends. Harley Cluxton III, a friend of mine and expert car collector, had managed to acquire the Shelby Cobra 'Super Snake' and was taking it to auction. One of two built, it was the only one in existence, the other having been destroyed (but immortalized by Bill Cosby's album *200 M.P.H.*). I was looking forward to seeing Harley and Carroll.

1966 Shelby Cobra Super Snake. (Courtesy *Motor Trend*)

SUPER SNAKES

In the winter of 2006, I was still pondering how to bring the Cobra Concept, GR1, or Shelby GT to market. The business problem with building low-volume supercars is not the cost of manufacturing the vehicle. Wealthy, car crazy, enthusiasts and collectors are more than willing to spend upwards of $200,000 to satisfy their desire to own and drive a beautiful, exotic, high-performance sports car. The problem is the up-front investment to design, engineer, and tool a car for production. These costs are nearly the same for a low-volume car as for a mass-produced vehicle, making the time to recoup the investment untenable from a business standpoint – no matter how high the vehicle price.

Somewhere in the back of my mind, I kept searching for solutions to this vexing problem that has frustrated car designers for years. One evening I was flipping through the TV channels and came across a movie I had not seen in years: Frankenheimer's 1966 *Grand Prix* starring James Garner. As I sat back and watched the movie, a strange thing happened. I can remember the exact scene where the camera zoomed in on a front three-quarter view of what appeared to be a replica Lotus F1 car (coincidently, these fake F1 cars were made by Shelby!). In that single frame of the movie, thoughts and questions started to race through my brain: "Wasn't that the first year Colin Chapman stressed the rear engine and transaxle by attaching the suspension to it?" As the camera zoomed in on the driver I thought, "Yes, and the engine was bolted directly to the monocoque tub. Too bad he had to poke a hole in the tub for the driver's head, ruining its torsional rigidity!" Then it hit me, "I bet a structural torque tube with the suspension attached to the engine and transaxle at each end could be stiffer, lighter and stronger." As the movie rolled on, I kept mulling the idea over in my head. "I bet an aluminum tube seven or eight inches in diameter would do the job." When Grand Prix was over, I grabbed paper and a pencil and started making calculations while sketching out chassis concepts. "Why won't this work?" "Why hasn't anyone done this before?"

Indeed, the idea of a torque tube, or 'backbone' structure was not new, Chapman used it on the Lotus Elan years before, and, as I learned later, the great Czechoslovakian engineer, Hans Ledwinka, developed the concept in the early 1900s. So too, Chapman had made the powertrain part of the chassis with his F1 car, but, to my knowledge, no one had ever produced a production car that used a torque tube, stressed powertrain, and energy absorbing structure to build a self-sufficient chassis that could support multiple body styles. Automobile architecture had evolved from the 'horse and carriage,' with the body mounted on a frame. For the 'horseless carriage,' inventors attached a motor to the frame with a body mounted to it. In 1934 automotive design evolution came to a fork in the road when the French firm Citroën introduced the 'unitized body,' where the body provided the chassis structure, to which the chassis components and powertrain were attached. Since that time, all vehicles have either been of body-on-frame, or unitized construction.

The idea that had flashed before me seemed too simple and not very profound, but it might partially solve the problem that had been vexing me. Such a 'Uni-Chassis,' as I called it, could be modular and flexible, allowing attachment of different bodies, while wheelbase could be changed merely by modifying the length of the torque tube. This would reduce both investment and engineering costs, as much of those costs would be spread over multiple models. Moreover, the movement to green hybrid and fully electric vehicles was in its infancy. Most early efforts stuffed the requisite batteries into a battery box that fitted into the nooks and crannies of an existing body structure. The body would be reinforced to support the weight of both the batteries and battery box. With Uni-Chassis, the battery box could be a structural part of the backbone, thus saving weight and cost.

Now, most people fall in love with their own ideas, and lose their objectivity. I needed the idea vetted. The Barrett-Jackson auction in Scottsdale was coming up in January, and I sought the opinion of those I respected: François Castaing and Carroll Shelby. My first stop was to visit my mentor at AMC and Chrysler, the great François Castaing. Visiting with him at his beautiful winter home, I presented my sketches. Yes, the idea would work; similar things had been done before. François pointed out that legendary race car designer, Adrian Newey, attached the front and rear chassis structures to a carbon fiber tub on his race cars.

Next stop would be to see Carroll at Barrett-Jackson. Uncharacteristically, it was cold and raining on Friday, January 19, when I pulled into the Barrett-Jackson guest lot. Carroll was staying warm in a motor home parked next to the B-J entrance tent. I knocked and Amy Boylan opened the door and invited me in. I had talked to Amy a year before when we were working on the Shelby GT, but we had never met. Carroll had hired Amy away from Mattel to run Shelby licensed products, and later asked her to run Shelby

American. Carroll was sitting up front in the motor home and we sat down to catch up on things. After a while, I pulled out my sketches and said, "Carroll, you've seen just about everything in the auto business. Have you ever seen a chassis like this?" I went on to explain the concept, and Carroll said, "No, I've never seen it. Have you shown it to anyone?"

"Just François," I replied.

"Don't show it to anyone until you get it patented." I agreed, and suggested that this chassis might help us finally do a new Shelby Cobra. Carroll agreed, and we decided that I would put together a proposal to build a 'proof-of-concept' prototype with Ford's help.

"We could even call it Super Snake II," I said.

That is when Amy chimed in: "A new Cobra would be great, but there may be a problem with the name," not letting on that there was another Super Snake in the works.

Later that day, Carroll and I walked through the display tent of featured classics. As we walked past the Super Snake, Shelby commented with a giggle, "We never could get the superchargers to stop throwing drive belts, and now it's the star of the show." Indeed, the story goes that Super Snake left Carroll stranded while on the "Turismos Vistadores" top-speed run from the California/Nevada border to Elko, Nevada. I spent the rest of the afternoon with Harley at his office and marvelous collection of historic race cars near the Scottsdale airport. Harley was in hyper-drive and could hardly wait for the auction the next day.

Shelby's Super Snake Cobra, CSX3015, started life as one of 19 (some say 22) 427 competition Cobras completed in 1965. Shelby had it modified in 1966 to be his personal car, if for no other reason than to see how fast it could go. The 427 was fitted with an Edelbrock cross-ram manifold, two Holley four-barrel carbs and most importantly two Paxton centrifugal superchargers. When tested by *Road & Track*, they declared it the "Cobra to end all Cobras." Carroll eventually sold the car to songwriter Jimmy Webb (*Wichita Lineman* among others), but it was eventually seized by the IRS for back taxes and auctioned in 1995

Super Snake with twin, belt-driven Paxton superchargers. (Courtesy *Motor Trend*)

for $375,000. Harley never told me how he came to own the car, as he always enjoyed keeping his acquisition of significant cars mysterious!

Saturday, Harley gave us passes to join him and his wife, Colette, in the skybox to watch the auction that afternoon. As the hour approached for Super Snake, he excused himself. Moments later Harley rumbled up on stage with Super Snake, to the cheers of a packed house under the big tent. Matt Stone was announcing the show to hundreds of thousands more watching at home, as Barrett-Jackson President, Steve Davis, introduced the car and asked Carroll to say a few words to the crowd. Carroll was uncharacteristically brief, "I got stopped doing 190mph in Nevada in it … did kill a buzzard in it … it's one of a kind … there's only one … it's a special car … it would do just a little over three seconds to 60, 45 years ago." Less than seven minutes later, to a standing ovation, Ron Pratte bought Super Snake for $5.5 million. It was the highest price ever paid for an American car at a public auction! Up on stage, Carroll was all smiles.

It took a while for the buzz to subside and the crowd to sit down and allow the auction to continue. Harley rejoined us in the skybox, his mind racing. "Let's buy a car for Colette. How about that Cord, what's it worth?" Colette tried to settle him down. Soon Carroll and Cleo joined us. He too was abuzz. Carroll wanted to know if the price paid was higher than that of any Ferrari. After years of beating him on the track, Carroll was still competing with his nemesis! Later that evening Harley invited me and SVT Chief Engineer, Jamal Hameedi, to dinner at a charming Italian restaurant. The celebration lasted late into the night, capped by rounds of Limoncello.

When I returned to Detroit, I took Carroll's advice and, on January 6, 2007, filed a provisional patent along with a trademark for 'Uni-Chassis.' I also went to see Manfred Rumpel, who upon retiring had started "Advanced Vehicle Technologies LLC" in office space rented from Technosports. We agreed to start on the proof-of-concept chassis design. By March, I had a draft proposal, and went to LA to review it with Carroll.

One of the advantages at ASC was our design office in Huntington Beach. I usually visited the office once a month, to meet with clients on the West Coast. I never missed a chance to see Carroll. When I was still working for Ford, we would meet at Shelby's office in LA on Olympic Blvd. Later on we'd meet at the Bel Air Club, where Shelby would always order a burger and Captain Morgan Rum with Diet Coke (Aaron Shelby tells me that Carroll actually preferred Diet Cherry Dr Pepper when it was available – it figures that Shelby was 'a Pepper,' whose tagline was "one of a kind").

"Don't worry," he said, "The doctor says I can have a drink once in a while."

Other times we would meet at his office in Gardena where Shelby's Goodyear tire distribution center had been established, or at a Tex-Mex restaurant he liked on Rosecrans near the 405 freeway.

On this occasion, we met at his office. The draft described the benefits of Uni-Chassis for both specialty vehicles, and hybrid and electric vehicles that would be coming in the future. It proposed building 'Super Snake II,' a Uni-Chassis proof-of-concept vehicle using an original aluminum 427 Shelby Cobra body. This would take advantage of the considerable publicity generated by the $5.5M sale of the original Super Snake Cobra at Barrett-Jackson, while laying the

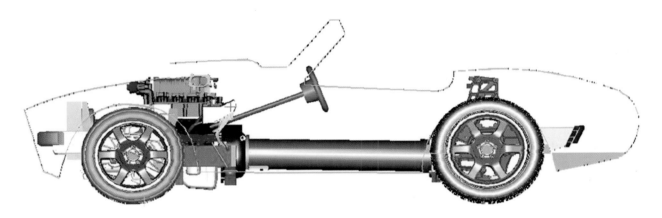

Preliminary Shelby Cobra Super Snake II package.

groundwork for a Ford-based aftermarket Uni-Chassis, and enable Carroll Shelby to do one last Cobra to cement his legacy. The proof-of-concept vehicle would utilize an enhanced Ford GT motor (650hp), a Ricardo transaxle, and Ford GT suspension, steering, brake, and cooling systems. Super Snake II would be used for Uni-Chassis development and verification of a preliminary Cobra II chassis (dimensionally identical, except for the wheelbase), along with introduction of the Uni-Chassis at SEMA and road tests with enthusiast publications. We agreed on the framework, but decided to wait until the preliminary design was complete before approaching Ford. Shelby also suggested hedging our bets on the 'Super Snake II' name with 'C/S Series II,' as they were planning to offer Super Snake conversions of Shelby GT500s the following year.

By June, the Uni-Chassis design was complete, and I met with Carroll again. We were ready to go and I scheduled meetings with J Mays and Derrick Kuzak on June 8 at Ford's Product Development Center in Dearborn. I met with Mays to let him know that there might be a new opportunity for developing specialty cars. More important was the meeting with Derrick. The proposal was to have my Super Snake Cobra chassis ready for SEMA in November, and a show car of Mays' design ready for the Detroit Auto Show in January. To accomplish that in such a short time frame, we needed Ford to authorize Ford GT suppliers to provide parts off existing tools, provide additional data, and supply one set of select parts for the proof-of-concept build.

If successful, production opportunities were: 1 – short-term, a limited edition run of Shelby C/S Series II; 2 – mid-term, an aftermarket rolling chassis to be sold by Ford Racing; 3 – long-term, various Shelby derivatives; and 4 – a future flexible low investment architecture for hybrids and electrics.

The Uni-Chassis idea intrigued Derrick. The potential for a new Cobra was, of course, interesting, but he saw Uni-Chassis as a potential technology that would help Ford achieve aggressive fuel economy standards that the government was considering. Fuel economy requirements would gradually increase to over 50 miles per gallon by 2025.

"We have a roadmap that will meet proposed regulations through 2020, but we don't yet know how to achieve the 2025 fuel economy goal," explained Derrick. "We know that we will need more battery electric and plug-in hybrid vehicles, but will need to develop new technologies and dedicated architectures to make them affordable." Derrick saw the proposal as an inexpensive way to explore a potential new technology. He agreed to authorize suppliers, supply data, and figure out a way to provide the required parts for Super Snake. Kuzak asked me to meet with Hermann Salenbauch, who was now heading SVT, to present the plan and work out the details.

We put together a more detailed presentation for Hermann and his staff on June 11. Using the Ford GT financials as a bench mark, the C/S Series II could be produced in limited volumes for significantly less than

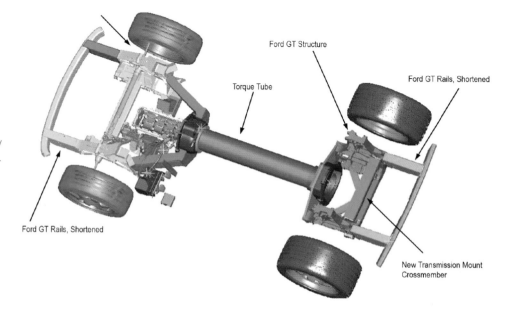

Uni-Chassis design for Shelby Cobra Super Snake II.

SUPER SNAKES

Stripped donor Ford GT, and ...

... Ford GT space frame.

the Ford GT. Variable cost would drop from $83,523 to $68,928, tooling from $51.5M to $27M and total program cost from $177M to $72M. The investment was still relatively high for a low-volume vehicle, but could be profitable for a series of production

First Shelby GT500 at 2008 Barrett-Jackson raised $550,000 for Juvenile Diabetes. (Courtesy Ford Motor Company)

J Mays and Carroll Shelby, NY introduction of the 2008 Shelby GT500KR: 40th Anniversary model. (Courtesy Ford Motor Company)

SUPER SNAKES

Original 1968 Shelby GT500KR: "King of the Road." (Courtesy Ford Motor Company)

2007 GT500 Super Snake. (Courtesy Shelby American)

derivatives. Hermann and his team agreed to support the development of the proof-of-concept Super Snake II. The most difficult part would be obtaining the long list of required parts, since the Ford GT was no longer in production.

Things always seem to take longer than expected, and sourcing Ford GT parts was no exception. It was eventually decided that the easiest way to obtain the required parts was to use a donor vehicle. Prototype Ford GT VIN# 1FAFP90S05Y400036 was scheduled for the scrapyard, but, having completed durability testing, became the donor. It took until June 2008 to finalize the purchase agreement, take delivery and start disassembly of the prototype for the necessary parts. I also managed to purchase the last Ford GT frame from Thyssen-Krupp, to modify for the front and rear structures of Uni-Chassis.

Terlingua Racing Team old & new. (Courtesy Shelby American)

THE LAST SHELBY COBRA

While the proof-of-concept was proceeding slowly, Shelby American was pumping out hot cars. In April of 2007 at the NY Auto Show, Ford and Shelby introduced the 2008 Shelby GT500KR. Shelby GT500s built at Flat Rock would be sent to Shelby American for upgrade to the 40th Anniversary KR (King of the Road) model with 540 horsepower, carbon fiber hood and splitter – along with a host of other performance modifications made by Shelby in co-operation with SVT. Shelby always had a story about how particular cars were named, from the original Cobra to the GT500, and the KR was no different:

"In 1968, I found out that Chevrolet was going to build a King of the Road Corvette. So immediately I called my lawyer in Washington and asked, if 'King of the Road' was taken.

He said, "I'll let you know by tomorrow morning."

I replied, "I'll have another lawyer by tomorrow morning! I want to know in an hour!"

And so the 1968 GT500KR was born. In 2008, 1011 KRs were built, with another 712 following in 2009.

To augment the Shelby American built Shelby GTs and GT500KRs, customers could bring their GT500s to Shelby American to be upgraded into Super Snakes with power options of over 725 horsepower. Shelby American even upgraded Mustang V-6s to a Terlingua Racing Team package, with optional Paxton/Vortech centrifugal blower that produced 375 horsepower. Things were going so well that Amy convinced Carroll to acquire a private jet to ease the burden of transporting Carroll between his many appearances across the country. Carroll, of course, called his old buddy Edsel, and bought the Falcon jet he had for sale!

As 2007 came to a close, things were looking good: The Cobra Super Snake proof-of-concept vehicle was proceeding, and customer orders for GT500 Super Snake upgrades were coming in. On November 8 I attended a sold out tribute dinner to Carroll Shelby at the Petersen Museum. Seated at the table in front of us were Carroll's children and grandchildren. The moderator asked Carroll questions, and, as always, Shelby mesmerized the crowd with his stories. I remember his closing remark: "Always try and be as good as your dog thinks you are!" Carroll laughed, and so did the crowd, rising for a standing ovation as the evening concluded. Indeed 2007 had been a great year, but things would be different in 2008.

2007 Tribute "Pit Pass."

Chapter 10
INTERESTING INTERESTS

The Great Recession started to take its toll on Shelby American. It was a mini version of the events that had led to Carroll quitting the car business back in the early '70s, and setting off to new adventures in Southern Africa. Back then, government safety and emission regulations put the Cobra out of production, manufacture of Shelby Mustangs moved in-house to Ford, and Henry Ford II suspended the Ford Total Performance program. This time, production of Shelby GTs at Shelby American in Las Vegas, including the introduction of the Barrett-Jackson Limited Edition Shelby GT, was starting to ramp down. The Shelby GT500KR, scheduled to start production in spring of 2008, was delayed due to quality issues primarily associated with the carbon fiber hood. KR deliveries did not begin until the fall, and the initial backlog of orders began to dwindle as the financial crisis worsened. The delays created cash flow difficulties at Shelby American and angst between Shelby and SVT. To alleviate the situation, plans were made to increase production of 2009 GT500KR models with 175 vehicles earmarked for customers around the world. By 2009, Shelby American was losing money. Carroll told me, "I need to keep my people employed." In fact, I know that, during the recession, Shelby was putting a couple of million dollars a year into the business to keep it going, and the private jet was long gone.

I was having similar difficulties at ASC. With the demise of the Chevrolet SSR, ASC needed to go through a Chapter 11 bankruptcy in order to offer a 'clean' sale to new investors. A Los Angeles based private equity company (that also happened to own Saleen), purchased ASC. For ASC, new business opportunities were drying up as the OE manufacturers cut back on product plans. This forced ASC to scale back operations. In the meantime, despite an infusion of more than $50M in Saleen, the company was hemorrhaging cash, heavily in debt, and up to its eyeballs with lawsuits. Steve Saleen retired, and the owners asked the ASC management team to try to salvage the Saleen business. Paul Wilber decided to place a big bet on the 2008 NY Auto Show, hoping to sell new ASC programs to the OEs, and more Saleen cars to customers. ASC/Saleen introduced new concepts at the show: a new Mustang with panoramic sliding glass roof, Ford F-150 with power folding tonneau, and a concept Saleen S5 supercar designed to draw a crowd. I knew the S5 would never make it to production, but was proud of the ASC design, led by Dave Byron. We even brought Dan Gurney to New York to introduce a limited edition Gurney Mustang, a tribute to the 1969 Trans-Am car prepared by Shelby American and raced by Dan.

Just after the NY show, US car sales collapsed. To make matters worse, Saleen had taken delivery of more than a hundred 2008 Mustangs for conversion, just

Saleen S5 'Raptor' Concept. (Courtesy Dave Byron)

Dan Gurney and Gurney Mustang. (Courtesy Dave Byron)

as the 2009 models were coming out. With financial disaster looming, Paul and most of the management team left later that year, and joined Aptera, a three-wheeled electric car start-up. I remained at ASC to try to stabilize the business and hopefully avoid an almost certain Chapter 7 closing of Saleen. Fortunately, Bert Boeckman, the largest Ford dealer in the world, and an old friend, helped me out. When I explained the problem of the excess consigned vehicles, he said, "Chris, I'll have trucks in Michigan this weekend and bring them back to California."

Bert had every reason to be upset, but understood the situation, and took charge without complaining or asking for a cent. No wonder Galpin Ford has been so successful! I could now try to keep as many people employed as possible by eliminating a layer of management and scaling back production to a more realistic break-even point, with an eye toward selling the company through an asset sale in hopes the new owner would retain our dedicated employees.

By the end of 2008, Saleen was ready for sale by the investment bankers. I stayed on working for several months without taking a paycheck. We successfully sold Saleen in February of 2009, and ASC became healthy enough to survive. In the depths of the recession, as most Americans watched half their life savings disappear, I too hunkered down. I put construction of Super Snake II on hold, and spent the rest of the year taking the Ford GT prototype apart, piece by piece, so that we would have all the donor parts available when fabrication resumed. This turned out to be a great learning experience for me. I had been heavily involved in the design of the Ford GT, and we obsessed about weight, Cg height and weight distribution during development. For example, we redesigned the spaceframe multiple times to eliminate every excess ounce of weight. With the compressed time frame, smaller components did not receive the same scrutiny. As I took the GT apart, I found areas where 100lb, or more, could have been saved – from the heavy steel muffler bracket, to the solid urethane 'Gurney' flap on the rear fascia and heavy rocker splitters. The teardown also inspired ideas to simplify and save weight for both the cooling and oil supply systems on the Super Snake II.

While discussing the doldrums of 2008 and 2009, perhaps this is a good time to reflect on the other aspects of Carroll Shelby's persona. I often think that had Shelby been a decade or two younger, the Dos Equis beer ad campaign might have portrayed him as 'The Most Interesting Man in the World.' Carroll was much more than a retired race car driver, race team owner, and builder of fast cars. I wish I had the journalistic skills to capture the essence of Carroll's spirit. The best I can do is provide examples of his curiosity, interests, storytelling, humor, connections, temperament, stubbornness, vanity, charm, philanthropy, controversies, wisdom, and friendship.

Curiosity

It struck me as unusual that a man in his 70s and 80s would continue to be curious and interested in exploring a wide variety of subjects and technologies, but Carroll was always working on new stuff:

OX2

I first noted Carroll's curious streak the day after we introduced Daisy. Carroll asked me if I could give him a ride from Cobo Hall to the Lafayette Hotel, where he was scheduled to give a speech at *Ward's AutoWorld* luncheon to announce the magazine's 'Ten Best Engines.' Carroll took the podium and began telling stories about engine development in the racing days. He congratulated the award winners and, to my surprise, concluded his speech with, "If you really want to learn about a revolutionary engine technology, go to www.OX2engine.com." After the luncheon, I asked Carroll what the OX2 engine was.

"It's a revolutionary engine with pistons arranged like the barrel of a pistol. It has only three moving parts, is small and light and produces gobs of low-end torque. It will run on hydrogen, gasoline – anything but peanut butter." Curiosity aroused, I checked it out the next day. It turned out that the inventor of OX2 was Steven Charles Manthey from Australia. Carroll and 'Pete' Petersen (of Petersen Publishing fame) along with others, invested in the creation of Advanced Engine Technology Inc to develop the engine. To understand it, imagine the revolver cylinder of a six-shooter handgun. The OX2 was an eight-shooter and, instead of bullets, pistons moved fore and aft the 'pistol cylinder' according to the profile of a cam ring. This created a four-stroke cycle where two cylinders fired on each rotation of the revolver cylinder (ie 'engine block') as shown in the exploded view of the OX2 engine.

As head of Engineering at Ford and Chrysler, ideas for new engines seemed to come across my desk at least once a week. Most tried to rearrange the mechanical systems of a four-stroke engine to achieve some perceived advantages. While the proposed advantages

INTERESTING INTERESTS

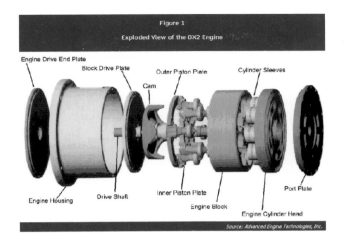

Exploded view of OX2 engine.
(Courtesy Shelby American)

Shelby & John Luft with OX2.
(Courtesy Shelby American)

sometimes existed, the potential disadvantages were either underestimated or not even recognized by the inventors. Most of these proposals ended up in my circular file, as even if they had merit, the cost, time and risk associated with solving all the hidden development issues typically doomed such projects to failure. OX2 looked like one of the better ideas to come around, but I perceived some critical issues as well, and did not pursue the matter any further.

Shortly after I turned down the offer to run Shelby American, Carroll called and asked if maybe I could do some consulting for him. Sure enough, he asked if I could take a look at the OX2 engine and help with some development issues. Carroll told me that he had vetted OX2 with a retired Ford Director of Powertrain Research, and now needed someone to provide advice to the team. Before going out to LA, I contacted the retired executive to obtain some background information. To my surprise, he was not nearly as enthusiastic as Carroll had intimated. His initial report accepted the advantages, but pointed out serious developmental issues as well. Ever the entrepreneur, Carroll had heard the good, but discounted the bad.

I flew out to LA and met with John Luft, who was running AET at Shelby's Gardena facility. John explained the background of OX2, and introduced me to his small team of perhaps six guys. Their 'office' was in a corner out back in the shop area where they machined Shelby 427 engines. I spent the rest of the day with the team, understanding the design, reviewing data, and looking at failed parts that had come off the prototype dyno engine. The team was an enthusiastic bunch, and I was impressed with what they had accomplished. One of the main issues, however, was that the seals that ran across the intake and exhaust port were failing in a matter of hours, if not minutes. Moreover, while low speed torque was prodigious for a 2.0L engine, very high piston speeds limited engine rpm. The idea had been to mate OX2 with a 30kW generator to create a lightweight (under 300lb) stationary generator package. I made some suggestions on the seal design, but also asked the team to take a step back and compare their stationary generator package to that of conventional 2.0L four-cylinder engine with 2:1 reduction gear. OX2's low-speed torque advantage was essentially due to the mechanical leverage built into the design. As a gut check, the team needed to know whether the product had a competitive advantage over conventional technology. I reported my observations to John Luft and Carroll, and told them they had a "long row to hoe." John said they had established funding, and would do their best to succeed before the money ran out. I returned to Detroit, but did not have the heart to charge for the consultation.

On my next visit, I stopped in to see the team. To my pleasant surprise, the latest OX2 design was running on the dyno. Durability had improved, but they still had to address myriad issues. To my chagrin, they had not bothered to calculate how the OX2 generator set would compare with conventional technology – an ominous sign.

I did not see OX2 again until sometime in 2011. While walking with Carroll around the Gardena garage where Carroll kept his collection of significant cars and remnants from his many other projects, we came across the OX2 generator set standing in an open wood crate frame wrapped in cellophane near the shipping dock. "What happened?" I asked.

"When the investor money ran out, I kept the project going for a while. We took it as far as we could," replied Carroll without regret. "It's time for a large company with heavy resources to finish the job."

The decade-long experiment had ended, and I never learned what became of the OX2. As I write this, home generators have become very popular. The latest Generac 30kW systems, powered by 1.5L four-cylinder engines, weigh less than 400lb with cooling system and enclosure. AET had identified the right market, but taken on too difficult a development challenge.

Electric Vehicles
In 2003, while working with Carroll on Daisy, he mentioned that he had experimented with electric motorcycles. We were walking though the Gardena garage, and Carroll pointed to the far corner: "I plugged it in to charge it. As I was walking away, the damn thing blew up and caught fire! Damn near killed me. That was the end of my experimenting with electric vehicles."

A few years later, Carroll gave me a call.

"Chris, the boys from Tesla just brought over a prototype roadster for me to drive. It's a real sport car and a hoot to drive. You should check it out."

"Naw," I replied, "It's just a Lotus with electric motor and a bunch of laptop batteries taped together."

In the intervening years, Tesla tried to recruit me a couple of times, but I turned down their requests for interview – I don't believe in working for companies whose business models are untenable.

Five years later, I had to eat my words. As guest judge for *Motor Trend*, despite reservations about the business case, I voted for the Tesla Model S as 2013 Car of the Year. At the awards ceremony in New York, I had the opportunity to talk with Elon Musk. I congratulated him, and put in a good word for Hubert Mees, who had designed the Ford GT chassis and now worked for Tesla. To my surprise, Elon was well aware of Hubert and thought highly of his work. Never very tactful, I blurted out, "I love the Model S, but don't see how you can continue to sell a car for less than it costs to make!"

"Don't worry, I'll handle that," he replied with a twinkle in his eye. The next week, Tesla raised prices. Nevertheless, despite heavy government incentives, Tesla has struggled to turn a profit.

Basalt and Carbon Fiber
On one of my visits, Carroll mentioned that he was experimenting with a volcanic fiber material, and had invested in a Russian based company that was producing volcanic fiber: "It's as strong as carbon fiber, and has much higher heat resistance. We've got it on test on with brake components and tanks on a fleet of commercial trucks."

I didn't think much about it at the time, but a few months later, Carroll had one of his guys give me a call, "Carroll wants me to make a 427 Cobra body for you out of volcanic fiber matt, instead of fiberglass." Carroll and I had decided not to build Super Snake II with an aluminum body, since we knew modifications would be required. Instead, we decided to build the prototype with a fiberglass body, and then retrofit with carbon fiber.

"If the body is as strong and light as carbon fiber, I'd be happy to use it," I replied. I decided to do some research once I got off the phone. Indeed, basalt fibers were made from melting volcanic rock, and exhibited high heat resistance (good for brake pads and clutch material), but the stiffness and strength properties were only somewhat better than the highest grade glass fiber. I sent Carroll a note explaining what I'd learned, but stated I would still be happy to use it on Super Snake II anyway. Carroll canceled the basalt fiber body, and we never discussed basalt again. I have since forgotten the name of the company, but note that production of basalt and usage in the automotive and building industries has grown substantially over the ensuing years.

Carroll also talked about a new carbon fiber that was stronger, "weighed one-third less, and cost half as much." I vaguely recall chasing down a coiled carbon fiber technology, but cannot remember if Carroll instigated the search or if, indeed, this was what Shelby was talking about. I do remember my search coming to a dead end, as the technology was sold to a Mexican company. In any event, coiled carbon fiber and coiled carbon fiber nanotube technology has advanced, although I do not think any results have achieved the performance characteristics Carroll was touting.

The point is simply that Shelby, in his 80s, was still curious and searching for technologies that were different and showed potential for the industry. As he said, he had "high hopes" for OX2 and the new carbon fiber. Carroll did not mind if his ventures didn't pan out. To him, the chase was more exciting than the final result.

Interests
Flying
Shelby joined the Army in 1941 at the age of 18, but insisted on being in the Army Air Force because he wanted to fly. As America's involvement in the war escalated, a change in the rules allowed for Carroll

to become a pilot even though he had not been a cadet. Hoping to become a fighter pilot and see action in the war, Shelby was instead assigned to fly trainee bombardiers and navigators around Texas and Colorado. Although he hated the boring job, Carroll's love of flying stayed with him, and Carroll's time in the Army Air Force put him in contact with the likes of Chuck Yeager and other flying heroes like P51 fighter pilot Bob Hoover. During the war, Bob was shot down and captured, only to escape, steal a Focke-wulf and fly to freedom in the Netherlands!

I did not learn of Carroll's love of flying and airplanes until he mentioned one day that he and Cleo had been up for their yearly visit at Baron Hilton's ranch, flying 'crop dusters.'

"Carroll, with your bad eyes, how do you keep a pilot's license?" I asked.

"Don't need one when you are flying over your own property," he replied. Now, I do not know if that is true, but Hilton's Flying-M Ranch near Reno Nevada had its own runway and covered 850 square miles! Cleo did get her flying license, and I can just imagine her, Carroll and Baron (four years younger than Carroll) barnstorming the countryside.

Carroll also went to the Reno Championship Air Races every year. After reading about the deadly crash in September of 2011 that took the lives of the pilot and ten spectators, I could not get ahold of Carroll. More than a week passed before I heard from him. "I was so depressed, I just couldn't call," he said, "The day before the crash, I was having dinner with a couple of the fellas, and they were saying that owners were hopping up their engines beyond the capabilities of the air frame – setting up vibrations in the tail." In fact, the ensuing investigation blamed improper modification to the P-51 Mustang, nicknamed the 'Galloping Ghost.' "I don't go to funerals any more – too depressing. All my friends are dying off." It was one of the rare times I witnessed Carroll in a 'down' mood.

Medicine

I doubt Carroll had a natural interest in medicine. His own physical circumstances brought the subject to him – the heart murmur as a child and subsequent maladies, the failure of his chicken farm, and his second significant crash while racing. Carroll was test-driving a Maserati 450S for the inaugural Riverside race in 1957, when he lost control and ran into an embankment. Carroll would describe in graphic detail how a plastic surgeon at the community hospital had saved his face.

"He stapled my nose to my forehead to reconstruct my face. It took 300 stitches [only 70, as reported by William Edgar, the Maserati owner's son, in 2012]." Hyperbole aside, Shelby had remained friends with the surgeon, whose name I can no longer recall.

More significant was Dr Alfredo Trento, who replaced Shelby's heart in 1990, and the Mendez brothers who transplanted Carroll's kidney donated by his son Michael in 1996. Doctors seemed to take a special interest in Shelby; after all, he was one of the longest-lived transplant recipients. Carroll was intrigued by their knowledge and education. From what I could tell, Carroll was a good patient and tried to take good care of himself. It seemed like he would have an appointment with at least one of his doctors every week. More importantly, Carroll was open to experimental procedures. On one of my visits, Carroll described how his doctor had ordered a special piece of equipment from Germany, "They strapped me down on the table, wrapped these bands around my legs, and then it would go bang, bang, bang and magnetically force the blood in the veins to help open new capillaries." From Carroll's description, it seems that he was going through an early form of what is now known as PEMF (pulsed electromagnetic field therapy), which is now FDA approved to improve micro-circulation, to increase capillary blood flow, promote healing of bone fractures, and treat arthritic inflammation, etc. Carroll also told me that his failing eyesight was due to macular degeneration, and that he was seeing a specialist regarding some experimental treatments, although I never learned exactly what treatments were being contemplated.

Kobe Beef

Everyone knew about Carroll's failed chicken farm, and I knew Carroll raised cattle, Shetland ponies, goats, chickens, and other assorted animals on his farm in East Texas. Shortly after the millennium, Carroll ventured into breeding African Tuli cattle as an experiment. His experience in South Africa led him to believe that they would thrive on Texas grass, despite the heat, humidity and parasites. Tuli meat purportedly rivaled the finest grades of Angus cattle, and Carroll owned nearly half the Tuli population in the United States.

I suppose Carroll was looking for another experiment when he told me, "I'm going to raise the first Kobe beef in this country." Not into animal husbandry, I could not imagine why Shelby would want to get into such a venture. As usual, there was an 'angle.' In 2009, the US banned the importation of real Japanese-bred Kobe beef. I was skeptical of the reported wonders of Kobe meat. My days in Ireland with DeLorean had jaded me when the chief engineer told me there was "nothing better than a good Irish steak." Trust me, good old American bred Black Angus will beat any Irish steak for flavor, tenderness, and texture. 'Real' Kobe beef, however, takes steak to an entirely higher level, as I experienced in my travels to Mazda, Mitsubishi, and Toyota. Apparently, Carroll planned on breeding Japanese Wagyu cattle, and raising them with the pampered care and feeding of the traditional Kobe breed. Reportedly, this domestic Kobe beef was selling for a $4 premium over top grade beef, and Shelby's ranch began producing 'Organic Shelby Kobe/Wagyu beef.'

Story Telling

Carroll lived an incredible life, full of varied experiences, and was a remarkable storyteller. He was not one to dwell, or live in the past, but magnificent stories would come out, in response either to a question, or during a new project. He told stories with unforgettable imagery, and a mix of hyperbole and Texas similes ("shit through a goose," "fart on a hot skillet," and "couldn't pull the string out of a dead cat's ass" come to mind). These stories seemed to fall into two categories: those that had been asked and answered thousands of times, and those that spontaneously came out in relation to the topic at hand. Of the former, most Shelby enthusiasts know the stories about the chicken farmer's failure, the bib overalls, the meetings with Iacocca to get $25K to build the Cobra, or with Henry Ford II providing brass badges ordering a win at Le Mans in 1966, along with the failures and successes at Le Mans. Carroll recited these stories nearly word for word each time the subject came up, but they seemed fresh every time Carroll told them. Other stories were reduced to short hand. "What happened at Le Mans in 1965?" I asked.

"Soft head bolts," was the reply. Indeed, a batch of improperly heat-treated bolts made their way into the Le Mans entries and stretched under the stress of racing. The resulting head gasket leaks took the 289 powered GT40s out of the race.

Regarding Ken Miles' fatal crash in the J-car at Riverside, the investigation into the cause of the accident was inconclusive. When I asked Carroll, he thought otherwise, "It was the bonded honeycomb aluminum panels … The chassis broke in half … Ken was a dear friend … Not a day goes by that I don't think about him."

Similarly, the story of Shelby's job interview with Enzo Ferrari is well known: when Carroll asked about remuneration, Ferrari answered that to drive for Ferrari was an "honor." Carroll also told me that he had stopped in to see Colin Chapman:

"The cheap bastard!" said Shelby, "when we finished discussing the possibilities of racing for Lotus, Colin asked me if I would mind taking a radiator to GM's Harrison Division in Lockport, NY, to see if I could get a better price." The discussion continued regarding Chapman. Despite being a brilliant engineer, his penchant for "designing in lightness" and cutting corners to save money, had cost the lives of many drivers. There had been rumors that Lotus had skipped the Magnaflux or X-ray inspections on critical suspension components.

Carroll's down-home demeanor and storytelling made him relatable to everyone, from everyday laborers to CEOs, politicians, and dignitaries. The friendships he made led to even more stories. For example, Shelby's first sales pitch to Lee Iacocca turned into a life-long friendship. He would tell me stories of how he and Lee would "pal around" Boca Raton in the early days, or how Lee convinced Carroll to turn down the Toyota Gulf-states distributorship because Detroit was going to drive the imports "back into the sea" (a quote Lee apparently borrowed from Henry Ford II – or perhaps it was the other way around). Carroll then convinced his racing buddy and African hunting partner, Tom Friedkin, to go after the distributorship, and it eventually became a multi-billion dollar business! Despite Lee's bad advice, Shelby and Iacocca remained friends and neighbors in Bel Air, and continued to get together socially.

It seemed to me that, as the years went on, Carroll's stories focused more on giving credit to those who had made success possible. Shelby would point out the contributions of a "bunch of hot rodders" and race drivers. In particular, had Shelby not "inherited" Phil Remington and Warren Olson when he took over the Lance Reventlow race shop, the racing successes of team Shelby might not have come to pass. Phil's innovations are too numerous to mention, but began with making the first Cobra fast and reliable, and continued with the development of the GT350, and most importantly development of the GT40s. In particular, the Mk IV's success in 1967 would not have come about without Phil's aerodynamic changes, quick-change brakes and "warm" cooling ducts (to keep brake rotors from cracking at the end of the Mulsanne straight). Shelby also picked talented racers whose skills went beyond driving. There was Pete Brock, without whom there never would have been a Daytona Coupe, Ken Miles whose development skills were invaluable in making the cars faster, and Dan Gurney who contributed with innovations like the 'Gurney flap' and winning race strategy with AJ Foyt at Le Mans.

Nor was Carroll's praise limited to his team. Carroll pointed out the contributions of Ford employees, especially those who were willing to get their hands dirty, like Klaus Arning and Bob Negsted, "Bob designed the 427 Cobra chassis and suspension using Ford's new computer programs. We couldn't have done it without him." Coincidently, Bob worked for me at Ford for a brief period as a chassis expert, before retiring. A talented but quiet engineer, he never told me of his exploits with Shelby on the Cobras and GT40s. Sadly, he passed away shortly after retiring. Carroll also enjoyed telling the story of Ford executives going to the Ford foundry one night, changing clothes, so they could 'rub cores' to create the 427 engine blocks (a method for thinning the sand cores to create thicker cylinders walls, allowing for increased bore and displacement).

In a few instances, Shelby would also castigate those who had done him wrong. I had long ago avoided bringing up the topic of the Series 1, as I had watched the program implode from afar. One day I was rooting around the Gardena shop studying old Cobra construction when I stumbled across a couple of Series 1 chassis up in the mezzanine. To my surprise, I found (as Coletti had observed earlier) that the chassis design was fundamentally sound. When I returned to Shelby's office, I complimented the chassis, and then Shelby went off, "It could have been a great car if John Rock hadn't been forced out, and we had the supercharged Oldsmobile V-8 ... I was sick with a bad kidney, and got roped into doing the project by Don Rager. Then Larry Winjet, who owned Venture Industries where the bodies were made, decided he wanted to invest and become part owner. I had to take the deal to keep the project going. We couldn't build cars because the body panels wouldn't fit, costs went up and production was delayed." I was not surprised by Carroll's tirade, because in that time frame I was having difficulty getting hardtop roofs from Venture for the 2002 Ford Thunderbird. "The Vice Chairman promised to come out and help," continued Carroll, "he would fly out to Vegas, shack up with his secretary for a week, promise fixes, and then leave." By April 2003, Venture declared bankruptcy, and Winjet was later sued for siphoning $314 million from Venture Industries! Shelby American

only produced 249 Series 1 models. Fortunately, Carroll was able to extricate Shelby American from the debacle. I never brought up the Series 1 again.

One final story that I have never heard repeated is fuzzy in my memory bank, but worth telling. It seems that Carroll was having lunch in a local restaurant after a race, presumably in Europe, with his Le Mans-winning co-driver, Roy Salvadori, and his wife. Zora Duntov,

Shelby Series 1 at the Las Vegas plant. (Courtesy Shelby American)

Corvette chief engineer and notorious womanizer, was visibly inebriated as he sauntered up to the booth and began harassing Roy's wife. A scuffle ensued, but I am not sure if it was Roy or Carroll who knocked Zora out. Perhaps this event motivated Carroll's desire to "kick the shit out of Corvette." Wish I had asked. Later, it was Zora who nixed the deal for GM to supply Carroll with engines for his 'sport' car.

To paraphrase *The Naked City*, there are thousands of stories in Shelby's life. Too many to recount by his friends and colleagues, but all are entertaining.

Humor

It is impossible to capture Carroll's seemingly ever-present humor. Every Texas phase, gesture, or inflection seemed to embody the behavior of a mischievous child that refused to grow old. Esteemed racing driver and journalist Denise McCluggage reminded me of Carroll's peculiar laugh that carried with him from his early days in racing. I had the honor of presenting Denise with the 'Spirit of Ford' award the year before Carroll received his from Edsel. Whenever Carroll enjoyed one of his own jokes, his shoulders would start bouncing up and down. If the joke was particularly good, he'd look down, tears might come to his eyes, and then under his breath you could hear a 'tee-hee-hee,' shoulders twitching in rhythm. Finally, he would break into a boisterous raspy laugh.

One morning, Carroll told me he didn't get much sleep the night before. "Cleo and I broke the bed," he

Shelby getting a good laugh out of *Pornography and the Law*. (Courtesy Bill Warner)

said with a twinkle in his eye and shoulders jiggling as he awaited my response. I did not take the bait.

Clearly, Carroll relished life and enjoyed all its humorous twists and turns. Jay Leno pointed out that every photo of Enzo Ferrari shows him scowling, while every Shelby photo shows that big Texas grin.

Connections

They say Carroll carried a small black journal with him everywhere that had his ideas, notes, and calendar with contact information. I never saw it, but I sure wanted to get my hands on Shelby's black flip phone. Whenever we met at his office, that flip phone would be right next to him on the desk. Carroll used it as his central command post. Calls would come in from the likes of Edsel Ford, Keith Crain, Jim Farley (then at Toyota), and other notables inside and outside the industry. I remember sitting in Carroll's office on Olympic Boulevard, when Edsel called to check on the status of a navy blue FIA Cobra that Carroll was having built for him. "I'll get right back to you," and then Shelby would call Fenimore, tell him to hurry things up, and call Edsel back with an update. Many of us would talk to Carroll at least once or twice a week. One can only imagine how many conversations he had in a single day, but it kept him sharp.

If you asked Carroll a question, or needed something, he would immediately look up a contact and get on the phone mid-conversation. It seemed as if he had everyone's number, and rarely did he have to leave

INTERESTING INTERESTS

a message. Shelby had said that these days he might have been diagnosed with ADHD, but I think he was just impatient and wanted things done fast. He seemed to enjoy keeping lots of balls in the air at one time, and didn't want to let time get away from him. One night, Bill Warner's phone rang and he recognized the drawl:

"Bill? Shelby here. How many of those damn Cooper Fords did I build?"

Bill answered, "Eight."

"Thanks," Carroll said.

That was the entire conversation.

Carroll definitely kept up with everything going on in the industry, and loved to be on the inside of the latest news and gossip. When the Toyota unintended-acceleration story broke in 2008, Carroll called the next day. "Chris, Toyota stock tanked. They'll get over this. Time to buy," he declared. I did. It was one of the few, if only times, I made money in the stock market. Normally, my modus operandi had been to buy high and sell low!

Another time I was sitting at a table at the Ford party in LA for the introduction of the 427 concept car where Shelby was also being honored. I struck up a conversation with the gentleman seated next to me. It turned out that he was a retired senator from Nevada, who'd been involved with allowing Carroll to use prison laborers to build Cobras. He had come to the party just to see his old friend.

Simply put, Carroll was connected like no one I have ever known.

Temperament

While Carroll was almost always jovial, he could have a temper. I experienced this twice. The first time, I annoyed him by asking a stupid question:

"Carroll, do you still have the molds for the Super Snake hood?"

"Hell no," he barked. "There were no molds; the guys just banged the scoops out of aluminum."

The second time was a bit more dramatic. I had heard a story from the investors of Saleen that they had approached a close friend of Carroll to find out if investing in Saleen was a good idea. The friend asked Carroll, who reportedly said, "Why invest in Saleen, when you can invest in Shelby?" I thought the story was amusing, but when I told it to Carroll, he jumped out of his chair and yelled, "That's a God Damn lie! Who told you that?" I told him that I thought it was just a joke, but it took Carroll a while to calm down. I learned to be a little more careful of how I relayed stories.

Stubbornness

Although I never witnessed it, I know Carroll had a stubborn streak. Not just to win or see something through, but also whenever his mind was truly set on something. My only evidence is the wildly successful Daytona Coupe designed by Pete Brock. Carroll chose to listen to the criticism of a Texas colleague, Benny Howard, an aircraft aerodynamicist who disagreed with Brock's 'Kamm' tail design. Decades later, Shelby,

Shelby with Peter Brock's 'Kamm' tail creation, the Daytona Coupe. (Courtesy David Friedman)

THE LAST SHELBY COBRA

only grudgingly, admitted that Brock's design was a success, but still griped that the Daytona needed a rear spoiler to create rear downforce.

Vanity

I am not a good judge of a man's looks – best leave that to the women. Judging by early photos of Carroll and the many women in his life, however, I thought that, in his younger years, Carroll was a handsome man, with

Shelby dressed in black. (Courtesy Shelby American)

Shelby, March 1965. (Courtesy David Friedman)

INTERESTING INTERESTS

Trademark bib overalls. (Courtesy Shelby American)

One time I was worried about Carroll, having not been able to reach him for several weeks. Finally, he answered the phone.

"Carroll, are you OK?" I inquired.

"Yep, the reception is not too good out on the ranch … I'm driving back in my Jeep … Cleo's my seeing eye … she tells me where to go!" he said with a laugh, as I heard Cleo chide him in the background. Changing the subject, "That 4-liter engine of yours is the most durable of all time. Got over 200,000 miles on it and doesn't burn a lick of oil."

"Carroll, I didn't design that engine, I was just involved when it was time to get it into production," I corrected him. In truth, François Castaing had challenged the Jeep engine guys to get 50hp/L out of the old 1950s Nash motor, so that they could get rid of the leaky 2.8L GM V-6 that AMC was putting in the Cherokee. My role, along with engine development chief, Rick Reuter, was to make sure that the engine would last. We created the toughest durability test in the industry: 1000 hours cycling between maximum torque and maximum horsepower engine speeds. When an engine failed at 997 hours just before production was scheduled, we went back and made changes until the engine passed muster.

Carroll loved his fans and customers. For example, at the Amelia Island Concours, Bill Warner witnessed a young man run up to Carroll and say, "Mr. Shelby … My dad had a 1967 green Shelby GT350 … with some crazy option. Tissue dispenser or something … I forget … Do you remember it?"

"Yes, son, I remember it," replied Shelby, as the young admirer left with a huge grin.

"Carroll, give me a break," said Bill.

He answered, "If he's happy, I'm happy."

Carroll could sell anything to anyone, and knew how to attract and bring people together. Without his charm, few of his extraordinary adventures would have been possible.

a big smile and wavy hair – in the fashion of James Garner. It surprised me that, as an octogenarian, he still considered such things. Part of it was, of course, his brand: Carroll had traded in his trademark bib overalls for a black hat and black shirts or sweaters, and he always knew how to pose for pictures. One day in his office, however, I must have been staring at his stomach as he leaned back in his chair. I had not said anything, but Carroll immediately commented, "My stomach is distended because of the drugs I have to take since the transplants." I didn't realize I had been staring, but was surprised that his appearance was something that Carroll would worry about.

Charm

They didn't call Carroll the 'snake charmer' for nothing, and it wasn't because he tamed the Cobra. Nor was it a surprise that Carroll could charm the ladies. What amazed me was Carroll's ability to sincerely charm everyone with his character. He could instantly put you at ease, and was always ready with a personal compliment. It took me a while to learn how to tone down Shelby's complements, otherwise the stories would take on a life of their own.

Philanthropy

While waiting in the hospital for a donor heart, Carroll befriended a young boy who was also waiting for a heart. After Carroll's surgery, he started the Carroll Shelby Foundation to help transplant recipients. Sometime later, Shelby told me he also started funding an automotive trade school in East Texas to help teens. As the role of the foundation expanded, the charter changed to include "providing financial support for children and medical professionals to help

Shelby the 'Snake Charmer' with Don Prudhomme. (Courtesy Shelby American)

overcome life-threatening health issues worldwide and promoting continuing educational development." In addition to the charity car auctions mentioned previously, the foundation sold Shelby memorabilia, and Carroll signed autographs for donations – among the other usual fundraising methods. Cleo used to get involved too, hawking foundation memorabilia at events like the Barrett-Jackson auctions.

On one of my visits to Gardena, I found Carroll sitting at a folding table at the south end of the warehouse. The foundation staff were lining up literally hundreds of items for him to sign – everything from the traditional Mustang and Cobra glove boxes to clothing, models, even guitars.

"Carroll, doesn't your hand cramp up doing all those signatures for hours on end?" I asked.

"Naw, I learned to sign with my shoulder, not my wrist," he replied. Perhaps that is the secret behind Shelby's distinctive swirling signature.

Carroll didn't limit his philanthropy just to his foundation. Whenever I would ask for support, he would send all manner of Shelby memorabilia for the charity auctions I supported. I must admit that I was the high bidder on several occasions – just couldn't let them get away! Ever the competitor, Carroll would invariably ask how much his donations had helped raise.

Controversies
Shelby was a high stakes gambler. Throughout his life, he placed big bets, sometimes winning, other times losing. The losses did not seem to bother him, but people that invested with him did not always

INTERESTING INTERESTS

Autographs for the Carroll Shelby Foundation.
(Courtesy Shelby American)

feel the same way. That might explain some of the controversies that touched Carroll's life, but there are too many stories to dismiss indiscretions, bad deals, and hurt feelings. I am sure some are true, but I never witnessed any of them. Certainly, Shelby was no saint, and not without flaws, but this book is about my first-hand dealings with Carroll.

I think I managed to meet Carroll at the best time of his life. If there had been wrong doings, it seemed that in his advancing years, whenever something or someone was wronged he would try to make it right. That did not always happen. I know of at least one instance where Carroll tried to bury the hatchet with someone he had wronged, but the wound was too deep to be forgiven.

I never had bad dealings with Carroll, and no money ever exchanged hands between us. To me, his word was his bond.

Wisdom

Lord knows Shelby made a lot of mistakes in his lifetime, and Carroll was the first to tell you all the things he'd never do again. The two biggest lessons were probably not to try and do a complete car without a manufacturer as a partner, and the second was not let anyone else try to do it using his name.

Even though Carroll is credited with designing the Cobra, in truth, he adapted and modified an existing, obsolete AC Ace, stuffed in a Ford V-8 and created a winning sport car. With new regulations, the ground up design of the Series 1 taught Shelby how difficult and expensive it was to design and build a production car, even with established suppliers. He learned the second lesson from people who had money to burn, but didn't know a thing about the business – or as my music teacher once told me, people who "wanted to play badly, and did!"

I saw Shelby keep his vow when a newly minted millionaire 'acquired' the rights to the AC Ace, and started spending heavily to build a modernized Cobra. He had won a $400M patent infringement judgement, wanted to get into the car business, and asked me to see if Carroll was interested in licensing his name. I doubted Carroll would be interested, but did tell Carroll that I would fill him in on my next visit. When I met with Shelby, he had already checked the guy out through his sources. The $400M was real.

"Tell him it takes at least $100M to build a car, and he doesn't know how difficult it is to design and build 3000 individual parts. I wish him the best, but no thanks," was Carroll's answer. "I get these all the time," he said as he tossed another offer he'd received for me to look at (which I still keep in my files). In fact, the guy lost his $400M when the judgement was appealed, and many suppliers ended up with unpaid bills.

Shelby had been burned by another deal with Unique Performance; they seemed like nice enough guys when I met with them and Carroll at SEMA. They, along with Chip Foose Camaros, built replica GT350s and 'Eleanor' Mustangs – until the Feds raided them for 'title washing.' Carroll and Chip both had a tough time washing their hands of that mess.

Carroll had learned what not to do, but that certainly didn't keep him from trying new things.

Friendship

Over the many years of his life, Carroll collected thousands of friends, and millions of admirers. For me, what started out as an admiring fan eventually grew into a much deeper friendship. We became close confidants after Daisy, but it wasn't until after I retired from Ford that I recognized how close we had become. Somewhere along the way, phone conversations with Carroll always ended with "love ya, Chris," and "Love ya, Carroll." When we got together in person, it was the same way, but with the addition of a bear hug. My father had passed away when I was 21, at the age of 75, and sentimental as it may sound, Carroll felt like a second father to me – sharing stories, opinions, and advice.

I was taking a break at the LA Auto Show when I sat down for a drink with Matt Stone, one of Carroll's favorite journalists. We got to talking about Carroll and suddenly he broke into his best Shelby impersonation: "Love ya, Matt!"

At that moment, it struck me how large Carroll's circle of close friends was – never did I think Carroll's words were insincere. In the hierarchy within Shelby's circle, family came first. He often told me how proud he was of his kids (Sharon, Michael, and Pat), and how independent and successful they were, along with the importance of his grandkids. It seemed to me that the circle then expanded to include trusted people like Joe Conway, his sister Anne's son-in-law, and long-time friends like Bill Neale, Baron Hilton, Bob Petersen, Tommy Friedkin and others he mentioned in the Bel Air community, like Walter Miller – I'm sure there were many more I didn't know about. Next there were the drivers, many of whom unfortunately were dying off, and his colleagues at Shelby's enterprises, some of whom had an on/off, love/hate relationship with Carroll.

I hope I fit somewhere in the next level of the hierarchy: real car people – racers, journalists, mechanics, engineers and knowledgeable enthusiasts. People like Edsel Ford, Keith Crain, Bill Warner, Denise McCluggage, William Jeanes, Barry Meguiar, Brock Yates, and David E Davis were probably at the top of this list, which included many others like Jim Farley, Neil Hannemann, John Fernandez, Matt Stone, Jean Jennings, John Coletti, John Clinard and too many more to name and remember (forgive me – wish I had that flip phone!). Carroll loved y'all, and Carroll loved his fans!

One thing I know for sure is that Carroll hated imposters, and had no time for them.

Two miserable years passed, and in the meantime, I have hopefully given the reader a better understanding of Shelby, the man. As the nation slowly crawled out of the recession, 2010 became a lot more interesting. Shelby American was hanging on, awaiting the arrival of the new 2010 Mustang, in the hope that it would stimulate the market, and I was looking forward to building Super Snake II.

Shelby with family members: son Patrick, grandson Aaron, and great grandchildren Larson and Pierce. (Courtesy Aaron Shelby)

Chapter 11
UNFINISHED BUSINESS

The malaise of the Great Recession began to lift as rays of hope emerged for 2010. General Motors and Chrysler had emerged from bankruptcy. Under the engaging leadership of new CEO Alan Mulally, Ford had managed to avoid bankruptcy, and was gaining momentum in the marketplace. To signal its newfound strength, Ford introduced the 2010 Mustang at the Los Angeles auto show on November 18, 2008. Shortly after the successful introduction of the Shelby GR1 Concept in 2005, George Saridakis was promoted to lead exterior designer for the new model. This design picked up on other Mustang heritage design clues, and was meant to jump start Mustang sales post-recession. Production of the 2010 model began on January 12, 2009.

For Carroll and Shelby American, this meant there would be new business upgrading customer cars, and the opportunity to create new Shelby models. First off was the 2010 Shelby GT500, introduced just before Carroll's 86th birthday at the North American International Auto Show in Detroit. With a bold new exterior, skunk stripes and an enhanced 540hp supercharged 5.4L V-8 from the KR, the coupes and convertibles were bound to be hits. Even though the GT500s would be built in-house at Ford's Flat Rock assembly plant, this meant new licensing revenue

Shelby and Fields introduce the 2010 Shelby GT500.
(Courtesy Ford Motor Company)

Shelby GT350 *Motor Trend* cover. (Courtesy *Motor Trend*)

to support Shelby's operations. Sales of GT500s had fallen from 8152 in 2007 to 3004 in 2009, and without Mustang GTs being built in Las Vegas, a cash infusion was badly needed to keep Shelby American afloat. The first 2010 Shelby GT500 sold at Barrett-Jackson in Palm Beach raised $150,000 for The Carroll Shelby Foundation. Production began in the spring of 2009, and although only 2000 GT500s had been planned, strong enthusiast demand drove total production to 4458 coupes and convertibles.

For Shelby American, however, things had to wait for the introduction of the 2011 Mustang, since major powertrain changes were coming: a new 5.0L V-8 for the Mustang GT, and an aluminum 550hp supercharged 5.4L V-8 that was 102lb lighter and utilized the spray bore technology we'd hoped to use years before. Any engine development for 2010 Shelby models would quickly become obsolete, so work at Shelby American focused on the forthcoming 2011 models. A product plan was put together that included a 660hp Shelby

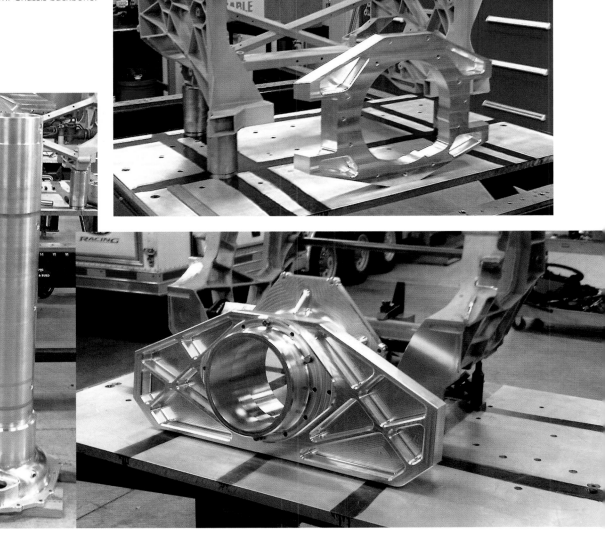

Clockwise from right
CNC front bulkhead;
CNC rear bulkhead;
Uni-Chassis backbone.

Ford GT motor fitted to front structure.

Ford GT transaxle fitted to rear structure.

UNFINISHED BUSINESS

Completed Uni-Chassis for SAE press release.

GT500 Super Snake, an optional 750hp Kenny Bell supercharger upgrade, and even an 800hp option. More prophetic, however (as you will see), was the decision to develop a Shelby GT350 to celebrate the 45th anniversary of the original. Shelby presented the concept at the Barrett-Jackson Gala in January of 2009. The press reacted positively to the concept of a 5.0L Whipple-supercharged Shelby, inspired by the original Shelby GT350 that ended Corvette's reign in SCCA racing. Featured in traditional white with blue stripes, Larry Wood of Mattel and longtime Shelby in-house designer and developer Vince Laviolette came up with a design that changed virtually every non-metallic panel on the car, creating an unmistakable GT350. First drives with the press did not occur until September of 2010, but the reviews were enthusiastic. My good friend, *Motor Trend*'s Jonny Lieberman, opined that the GT350 was the "fastest, best-handling and sweetest-sounding 'Stang we've ever laid hands on." With these 'post title' 2011 conversions, Shelby American was back in business!

Things were starting to get more interesting for me in 2010, as well. Shamel Rushwin, my co-conspirator in the move from DaimlerChrysler, asked me to consult on a possible acquisition of Volvo from Ford by Michael Dingman, billionaire investor and former lead director of the Ford board. While Dingman eventually lost out to Chinese manufacturer Geely in acquiring Volvo, this afforded me the opportunity to re-establish relationships with the Ford management team. In addition, Angus McKenzie, then editor-in-chief of *Motor Trend* invited me to be a guest judge during the magazine's annual Car of the Year testing in LA and the Mojave Desert. Not only did this allow me to stay current with all the latest cars and technology, it paved the way for more frequent visits with Carroll.

When I met with Carroll in August, I told him I was ready to start building Super Snake II. Carroll had been trying to get a carbon fiber 427 Cobra body built for me, but despite repeated promises from the supplier, it was never delivered. When I arrived in Gardena, the 427 molds were scattered around the shop floor.

"What happened?" I asked.

"My body supplier in Tijuana went bankrupt! The Mexican officials confiscated the molds, and wouldn't let me pick them up until I paid the salaries of every employee the owner had stiffed." Now Carroll was searching for another source. We decided to target completion of the chassis for the SAE (Society of Automotive Engineers) World Congress the following April, and Carroll promised do his best to get me a Cobra body for the display.

Back in Detroit, we really had to hustle to meet the April deadline. Detail drawings of every chassis component were completed, and then Technosports started fabricating the components. Most were CNC (Computer Numerically Control) machined aluminum billets intended to replicate production castings. The backbone, made from an 8in aluminum extrusion, was machined at a shop using a postwar Russian lathe big enough to handle the size. By December, all the machined components were complete, and we started jigging and assembling the front and rear structures, utilizing the corner castings and extrusions from the donor Ford GT frame. The backbone assembly was completed in January, 2011, and in February assemblies of the front, rear, and backbone structures were fitted around the stressed Ford GT motor and rear transaxle. By March 3, welding of all the structural components was finished and all the powertrain and suspension components had been bolted in place on the surface plate to complete the chassis. We submitted photographs just in time for the SAE press releases.

I was starting to get worried about receiving a Cobra body in time for the event. Carroll had not yet found a source to build new bodies, let alone a carbon fiber one. With the SAE show only a month away, Carroll grabbed one of the complete bodies that came with the shipment of molds from the supplier HST Automotive in Tijuana. March 4 was like Christmas morning when a huge crate arrived from California. Inside we found a beautifully painted 427 body in bumblebee yellow, with black carbon fiber racing stripes and racing roundels on the front fenders. I called Carroll and could not thank him enough for the surprise. I had assumed that the carbon fiber graphics were just fake appliqués that had been clear-coated, but later in the project; we discovered that the body was made entirely of carbon fiber. I doubt Carroll ever knew. Only years later did Gary Patterson (current Shelby American President) tell me that the body came from the showroom of the HST.

For the SAE Congress on April 11, we mounted the 427 Cobra body on a workbench so the body would appear to hover over the Uni-Chassis – an indication of our intentions to complete Super Snake II. Monday morning opened with a press conference where I explained the benefits of Uni-Chassis for specialty cars, hybrids, and electric vehicles, and thanked Carroll and Ford for their support. Interest was high, and I spent the next three days talking to journalists, entrepreneurs, and engineers from around the world. At the end of the Congress, Uni-Chassis received an SAE Automotive

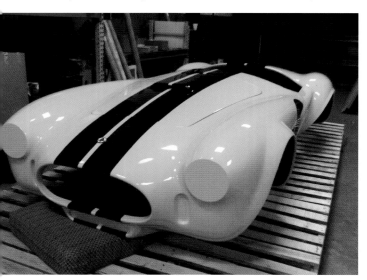

Carbon fiber racing stripes.

Shelby carbon fiber 427 body.

Engineering 2011 Tech Award. In the days that followed, articles appeared in newspapers, enthusiast magazines and technical journals. There were inquiries from both the US and abroad, especially China, the UK, and Turkey. Most were from start-ups interested in developing hybrid and electric vehicles, as well as specialty cars and even truck and bus manufacturers. While I entertained such inquiries, my focus was on completing Super Snake II so we could demonstrate the capabilities of Uni-Chassis.

Photo from the Uni-Chassis article by Frank Markus, May 2011 *Motor Trend*. (Courtesy *Motor Trend*)

On my next visit to see Carroll, I found him in his office, with copies of Uni-Chassis articles neatly stacked on the corner of his desk: *Automotive News*, *Autoweek*, *Ward's*, the *Detroit Free Press*, *Motor Trend*, and more. How he managed to keep track of all the industry news, I will never know, but Carroll was excited by the positive response. It was time to start mounting the body so that we could bring Super Snake II to life. We discussed the modifications that would be required to adapt the body to Uni-Chassis, including even wider fender flares to match the wider Ford GT track, and a new hood to clear the supercharger.

Back in Detroit, we built a wood seating buck to fit to the chassis and body, and work out interior ergonomics. Legroom was outstanding, the pedal box was large, and both the pedals and steering column were located on-center, contrary to the skewed seating position of classic Cobras. Unfortunately, the Sparco Ford GT seats were too wide. Not wanting to bother Carroll, I called Shelby American to see if they could provide a classic Cobra seat for fitment. Sure enough, the next day Carroll called.

"Why the hell do you want to use those crappy Cobra seats? Why don't you get some Recaros?" I explained that I couldn't find modern seats to fit, Carroll then digressed, "I should have listened to Negsted (and Peter Brock) and increased the wheelbase three inches. The Cobra wasn't perfect, we should have improved it. I tell everybody that the Ford GT is the best sport car in the world."

"Carroll, there you go buttering me up again," I responded, "There are plenty of things we could have done better on the Ford GT." Eventually, we decided to split the Cobra down the center and widen it by 9in. This eliminated the need for additional fender flares and allowed for comfortable fitment of the Ford GT seats, center console and McIntosh sub-woofer between the seats.

The Cobra body was adapted to the chassis by fabricating a body tub from bonded and riveted honeycomb aluminum panels. This was essentially the same construction technique used by the J-car and

Original Cobra body.

Cobra body after a 9-inch widening at the center.

1967 Mark IV Ford GT. Fortunately, great strides in the strength and durability of two-component epoxy adhesives have been made in the ensuing years. The body was bonded to the tub, and mounted with rubber isolators. Next came fitment of the remaining chassis components. We located the Ford GT dry sump reservoir in front of the passenger footwell, and fabricated a new oil cooler/filter/pressure relief system to replace the heavy engine-mounted arrangement of the Ford GT. The cooling system was modified to eliminate the Ford GT's two reservoirs, and a new top tank fabricated to provide separate cooling circuits for the radiator and intercooler. We even kept the Ford GT's anti-lock braking system, utilizing the space above the bell housing. In back, a new fuel tank was fabricated and fitted above the transaxle. The optional Ford GT transmission cooler was installed, and a battery located in the rear.

When I couldn't make it out to LA, I would send photos of progress on Super Snake II. I could tell Carroll's enthusiasm for the project was building. That summer he called: "Chris, you're going to get a phone call from Lenny Shabes. He wanted to do a show for Discovery documenting the build of one of my Cobras. I told him that's old news, and suggested he do a show on the build of Super Snake II." Indeed, Lenny did call. Lenny had produced *Battle of the Supercars* and many other TV documentaries. We discussed doing the show, but first he had to line up the Discovery Channel. We agreed to keep in touch as talks progressed.

The summer of 2011 got really interesting. Out of the blue, Carroll gave me a call and said, "Chris, I

Plywood seating buck.

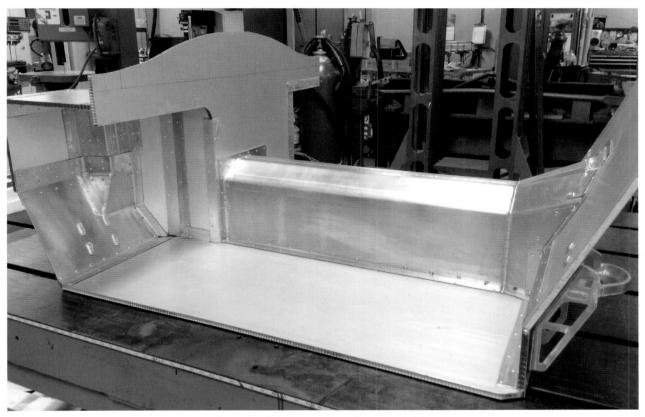

Bonded and riveted aluminum honeycomb tub.

want to build some cars. Put together a business plan for me." Over the next month, we began discussing a product plan. At first, we discussed developing fully certified sport cars, the dream being to produce a line of products to create a profitable Shelby American Supercar Company. Financial realities soon put an end to that dream. Even with the investment savings provided by Uni-Chassis, a modern Cobra roadster like Daisy would require a $77M investment, and we would need to sell more than 2100 cars at $125,000 to break even. Similarly, a modern GR1 priced at $200,000 needed a $92M investment and at least 1000 cars to break even. Regardless of investment, the realities were that Shelby American could not sell thousands of cars without an OE partnership.

Failing that, we looked at selling 'gliders.' These would be fully assembled vehicles with modern chassis components (ie, Uni-Chassis using either Ford GT or Corvette chassis suspension components) and a Graziano transaxle. Customers would then purchase and have the engine installed separately. Without all the safety and emissions requirements, investment dropped to between $2 and $4M, depending on model. We agreed to a five-year product plan that would start with production of a classic Cobra roadster using modern chassis componentry in the second year, followed by a 3in stretch, "for the basketball players," then a modern roadster and coupe in each successive year. Total planned volumes were 150, 75, 150 and 100 for the classic roadster, classic stretch, modern roadster and coupe, respectively priced at $150K, $150K, $175K and $200K. This 'crawl, walk, run' business strategy predicted impressive results. Cash flow would be positive in the second year and provide more than enough cash for investment in each subsequent model. Theoretically, at the end of the fifth year the company would have over $20M cash on hand! Lest anyone be tempted to borrow this business model, be forewarned: without the heritage and strength of the Shelby brand, neither the pricing, nor the projected sales volumes are necessarily achievable.

I mailed a copy of the final business plan to Carroll

in advance of my next visit. I was scheduled to go out to LA for *Motor Trend*'s 2012 Car of the Year testing, so we scheduled a get together on September 21. When I got to Shelby's office, we started with the usual hug, and Carroll introduced me to Joe Conway, jokingly (I think) referring to him as his "lawyer," when Joe was really his business manager. Waving the business plan at me he asked, "Is this plan going to work?" I explained that I tried to be very conservative with the investment numbers and volumes.

"If you think you can sell a hundred cars a year at these prices, it will start making money in the second year," I responded.

"Hell, we should be able to sell at least a hundred cars," replied Carroll. We then went on to discuss next steps, issues, and decisions that had to be made: meeting with GM or Ford on component supply, making a decision on Ford GT or Corvette C6 suspension components, sourcing of the body, perhaps Technosports supplying chassis and final assembly at Shelby American, etc. Personally, I could not assist in the investment, as I'd already spent over a quarter-million on the Super Snake II project. I told Carroll and Joe that I wasn't interested in making money on the project; I just wanted to help make it successful.

At the end of the discussion, Carroll said, "We can do this; I'll see if I can get help with the investment." With that, he grabbed his flip phone. "I'm calling Bill DenBeste. He's our largest Cobra dealer. I sold him the 427 engine business, and he's done a great job with it." Bill was not in, but Carroll said he would meet with him at SEMA, as well as other potential investors. Through the entire meeting, Joe Conway never said a word. After I left, it occurred to me that maybe Carroll was thinking about a licensing arrangement, or perhaps Joe was there so Carroll could get his input after the meeting. I never did learn what Joe thought of the plan.

As we left his office and walked through the shop, Carroll talked about his projects with Ford. He was excited that the upcoming 2013 GT500 would be capable of 200mph. He was scheduled to evaluate the car at Sebring in October. Gene Martindale, a key member of the Ford GT team, had been promoted to lead development engineer on the project, and Ford scheduled a film crew to record the testing with Carroll. As we walked around the shop, Carroll pointed out the little Dodge Omni GLH (Goes Like Hell) that he was fond of driving.

"I want do a light pocket rocket like that again," he commented. For whatever reason, the hot hatch

Carroll with prototype 2013 Shelby GT500 at Sebring, October 2011. (Courtesy Ford Motor Company)

never happened, even though the fantastic Fiesta ST was coming for the 2014 model year. I left, excited that a modern line of Shelby Cobras might be on the horizon.

In retrospect, I wonder what spurred Carroll into asking for a business plan. Back in the days of Daisy, Carroll had said that one of the things he wanted to do before he died was build another Cobra. As Carroll's grandson Aaron told me, he had always begged Carroll to do another car, only to be rebuffed and told, "You can't do it without the help of a manufacturer." Indeed, it surprised Aaron when Carroll started the Series 1 project – although GM had promised investment, engineering support and Cadillac Northstar engines and components at transfer prices when the project started. When GM pulled out, Carroll chose to carry on. After the fiasco with Winjet (another lawsuit against Winjet has been filed by the bankers as I write this), why would Carroll want to go it alone?

I think I know the answer. That same summer, Edsel Ford told Ford management that Carroll was getting on in years, and maybe "we should do something to help him, like licensing the GT350 name." Mark Fields told me that he had met with Carroll at a Mexican restaurant (probably the same one we used to meet

UNFINISHED BUSINESS

at) and broached the subject of securing Shelby's legacy with Ford. However, it was Carroll's old friend, Jim Farley, who got the assignment. Jim told me that after about three months of work with the lawyers, they got the GT350 licensing deal done, and all of Shelby's complicated business structures sorted out. To celebrate, Carroll suggested that Farley join him at the Aston Martin banquet at Pebble Beach in August, where he was to be honored for his 1959 Le Mans win, "We can celebrate with a glass of Champagne."

At the event, CEO Ulrich 'Uli' Bez (who was always talking, but never listening), got up and made a 45-minute speech about all his great accomplishments and the new transmission he had developed for Aston. Carroll was getting agitated. Finally, Uli remembered that he was supposed to introduce Carroll, and asked him to say a few words. Carroll got up and said, "Those Astons were junk, and we wouldn't have won if we hadn't put a Maserati transmission in them!" Uli was aghast, and Carroll leaned over to Farley and said, "Let's get out of here!"

Thanks to Edsel and Jim, the Ford/Shelby legacy

Carroll and Edsel at 2011 SEMA Show. (Courtesy Ford Motor Company)

THE LAST SHELBY COBRA

was settled, and finances for the estate secure. I suspect Carroll then wanted to give his dream of another Cobra one last shot, and secure the legacy of Shelby American in the process.

The SEMA show in Las Vegas was held the first week in November, and I anxiously awaited word regarding Carroll's success in lining up investment for the 'glider' program. Curiously, I didn't hear from Carroll, but I knew it was a busy time for him, so I planned to meet with him before the LA Auto Show in November. I arrived in LA just in time for the Ford party at the Belasco Theater on the eve of auto show press days. The theater was packed, loud and rocking with a light show and alternative rock band. Jim Farley and Carroll introduced the 2013 Shelby GT500. At the brief scrum, Carroll said, "I've always been asked. 'What is your favorite car?' and I've always replied, 'The next one.' Well, I'm taking that back tonight. This is my favorite car." As Carroll was guided off stage, I could see he was weak and tired as he approached.

"Chris, I'm sorry I didn't get to talk to investors at SEMA. Came down with pneumonia," he said in a whisper.

"Forget about that, let's get you someplace quiet where you can sit down," I replied. Cleo and I took him to a back room, got him some water and made him sit

Carroll and Jim Farley with 2013 Shelby GT500 at the Ford party. (Courtesy Ford Motor Company)

down. Lenny Shabes and a few others were there as well. We tried to talk some sense into him.

"Go home, and don't even think about coming to the press show tomorrow," I scolded.

"All right, all right," he responded, and after regaining his strength, Carroll went home with Cleo. Little did I know that Carroll had checked himself out of the hospital that day, against his doctor's advice, nor did he return to the hospital. The next morning at the Ford press conference, there was Carroll sitting in the stands with Cleo, so I joined them. Mark Fields introduced him during the unveiling of the 2013 Shelby GT500. Carroll stood up and waved to the applause of the crowd. Stubborn as always. It was the last time I would see him.

Back in Detroit, I would call Carroll at least twice a week to see how he was doing. At first, he seemed to be recovering. Conversations were brief, but despite my fears of pneumonia at his age, I was optimistic. Sometime in December, however, Cleo told me that Carroll had tripped on a rug, taken a fall and was confined to bed. Carroll answered the phone when I called during the holidays and seemed to be his usual gruff self. After wishing him a Merry Christmas, I asked to talk to Cleo. "What the hell you want to talk to Cleo for?" he growled, knowing that I would inquire how he was really doing. Cleo assured me he was on the mend. With Carroll's 89th birthday coming up, I bought two cards to send. I managed to find a 90th birthday card and marked it "do not open until 2013." Inside I wrote, "Damn it Carroll, I told you not to open it!"

Knowing that Carroll would be deluged with calls on his birthday, I waited a couple of days to give him a call.

"Did you get your birthday cards?"

"Yeah, I opened them both," he replied and we had a good laugh. When I signed off, his last words, of course were, "Love ya, Chris."

Sometime in late January or early February, Cleo checked Carroll into the UCLA Medical Center and said she had gotten his old doctor to come and straighten out his meds to get him back on track. I asked when I could come out and visit, but Cleo told me the doctors did not want Carroll to have visitors. Meantime, I was thinking about something Gary Patterson had told me on one of my visits to Shelby American. It seemed that they were still getting regular inquiries about the availability of Daisy. I had mentally assumed that we would always update the Daisy design if we did a new Cobra. It was then that I remembered that three fake Cobra Concepts had been made at the Saleen facility to be blown up in the movie *xXx: State of the Union*, starring Ice Cube. While the fake cars had been destroyed, the molds had been tossed out back in a scrap heap. I confirmed that they were still there. Hoping to cheer Carroll up in the hospital, I told Cleo to tell him I had found the Daisy molds, and we could start on the project as soon as he got better.

I continued to call twice a week to get an update on Carroll, and inquire about visiting. Every time, I would get a progress report and be told to wait. Later Barry Meguiar told me that he did manage to visit, and found a weakened Carroll weighing less than 100lb. "Probably better that you didn't see him, so you can remember him when he was well," said Barry. In April, at Carroll's request, his family CareFlited him to the Baylor hospital in Dallas. Cleo never told me, even though I continued to call regarding his health. On May 11, 2012, I awoke to the news that Carroll had passed away in Dallas. I had lost a dear friend and mentor, and went into a deep funk, never having had a chance to say goodbye.

Chapter 12
KISMET: DAISY COMES HOME

In the days and weeks following Carroll's death, my wife could tell I was upset and depressed. I could not sleep as my mind tried to recount every moment of my times with Carroll. News of Shelby's death went far beyond the racing and automotive communities. Press coverage was not only national, but also international. Carroll's amazing life story was on the national nightly news. Special sections appeared in all the major newspapers including *NY Times*, *USA Today* and even the *Wall Street Journal*. Magazines like *Autoweek* and *Motor Trend* put out special editions, and the staff at *Motor Trend* worked overtime to publish a book within two weeks. Every journalist and friend of Carroll had memories and stories to share – enough to fill a library.

The week after his passing, I received a call from Shelby American. A memorial tribute to Carroll was being planned for May 30, at the Petersen Museum; could I provide a list of names, particularly journalists and those in the industry, who were close to Carroll for the invitation list? I gladly complied, racking my mind to make sure I did not slight anyone. Shelby American, Ford, Shelby's family, and the Petersen

First Shelby Cobra, CSX2000 at the Petersen.
(Courtesy Tomy Hamon)

Museum, however, did the real work. It was amazing how in three short weeks such a magnificent event came together. For those who organized it, every Shelby fan cannot thank you enough!

I took an early flight to LA, checked into a B&B down the street from the Petersen, and walked over early to see if I could help in the preparations. It was early afternoon, and the parking structure was already filling with Shelby Cobras, Mustangs, GT40s and more. Inside the Museum, fantastic collections of Carroll's significant cars were on display, complete with signage: CSX2000 and 2001, Shelby Daytona Coupe, GT40 and 2005 Ford GT, King Cobra, and many more.

Shelby Cars on the Petersen rooftop. (Courtesy Tomy Hamon)

Opposite One of two *Motor Trend* Collector's Covers.
(Courtesy *Motor Trend*)

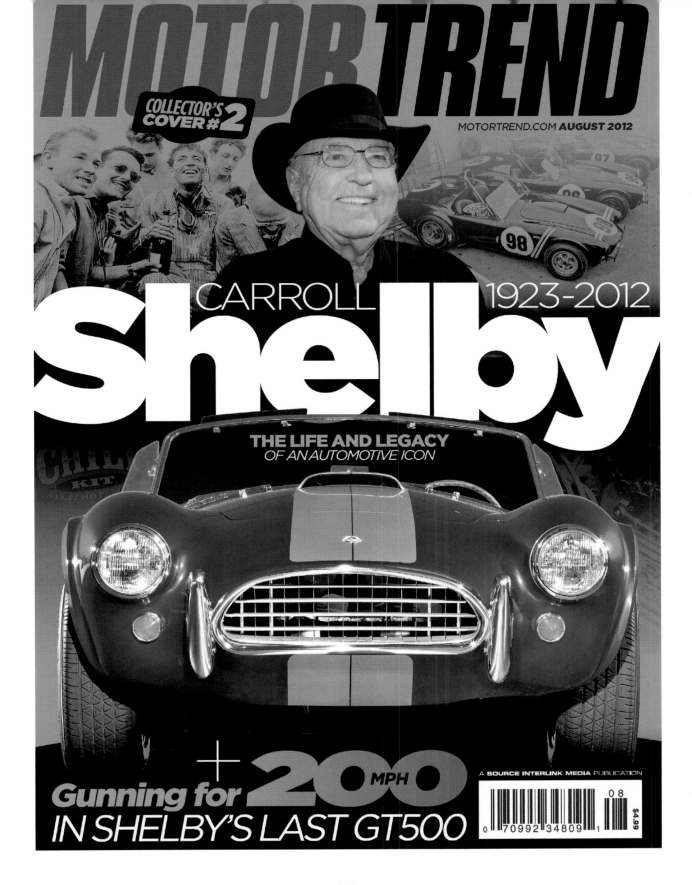

The 700+ invitees were treated to a strolling dinner among the rooftop display of Shelby cars. Chili was served, of course, and then the crowd was guided into the tented auditorium lined with classic photos of Carroll. In attendance were the many members of Carroll's family and circle of friends: Carroll's children and grandchildren, current and former employees, racers and fabricators, industry titans, Shelby owners and enthusiasts, journalists, publishers and historians – too many names to recite.

The memorial: *Carroll Shelby, A Life Remembered* was broadcast live on the web. The ceremony started with a historic video of Carroll's life. Jay Leno came up to the stage to act as moderator.

"People ask me how come I don't have any Ferraris? I tell them 'because Carroll Shelby told me he was a mean son-of-a-bitch' ... I never saw Enzo smile, and I never saw Carroll without a smile." Jay went on to intersperse his own stories throughout the night. Edsel Ford II described how he was first introduced to Carroll at Le Mans by his father in 1966: "a bigger than life personality," and pointed out that 1967 was the only time "an American team, in an American car, powered by an American engine, and driven by American drivers," won. "The automobile industry has lost an icon," he concluded, "and we've all lost a friend." Longtime friend Bill Neale recounted the first chili cook-off, and how Carroll fooled everybody by dressing up a 'professional woman' in a leather outfit, and introducing her as a descendant of an Aztec queen. Flying ace, Bob Hoover, explained that, because Carroll was at the top of his class in flight school, it was more important for him to be held back from combat, so as to teach others to be their best. Perhaps the most heart-rending moment was the testimony of Leah Smith, who, with Carroll's help, received a heart transplant at the age of two. It was Dan Gurney, however, who brought down the house. After asking all present and former employers who'd worked for their former leader to rise, he went on to give a great speech, concluding with, "It's a cliché to say they threw away the mold when they made him. In his case, that is a good thing, because whoever would try to make a new one would get sued by him for copyright infringement!"

Simultaneous with the tribute at the Petersen, celebrations were held in Las Vegas, headed by Shelby VP, Gary Davis, and at the Team Shelby Nationals, headed by Gary Patterson. At 6.55pm, the 'Rev Your Shelby' salute roared through the auditorium, as owners fired up the Shelbys outside and around the world – from the Shelby Automotive Technology Center at the North East Texas Community College to the Shelby club in Africa. Leno wrapped up the evening with a video, and the crowd began to mingle and carry on with stories. As I worked my way through the crowd, I congratulated Dan Gurney, and then whispered in Evi's ear, "How much of that did you write?"

"None of it," she replied with a wink. It was Dan's wit, but I'm sure Evi's PR polish from her days at Porsche racing helped make it so memorable. As it is too long for this chapter, I've appended a copy. The crowd slowly dispersed. On the way back to my room, I stopped in at an Irish Pub down the street, knowing that I would never get to sleep with all the images of Carroll in my head. There, in a booth I spotted Denise McCluggage, along with William Jeanes and Jim McCraw. I sat down for a drink and listened to the Shelby stories long into the night. It was a day that I will never forget.

In the aftermath of the tribute, the pain from losing a dear friend started to subside. Fearing I'd forget the many memories, and as a form of catharsis, I began writing this book, but stopped for the reasons mentioned in the prologue. I wish I hadn't. Meantime,

Dan Gurney at Shelby Tribute.

KISMET: DAISY COMES HOME

those at Shelby International and Shelby American were adjusting to the loss as well. They'd known this day would come, but it still shook them. Joe Conway took on the task of the finances and legal entanglements. In the midst of uncertainty, John Luft and Gary Patterson carried on with the five year plan that was already in the works. In January, while Carroll was ill, Shelby American had its first independent display at an auto show, with John Luft unveiling the 50th Anniversary cars, including the Shelby 500 Super Snake, available with over 800hp. I still have the gold 50th Anniversary lapel pin and brochures the team was handing out. Next up, the team met Carroll's 1000hp challenge just before his passing, and introduced the 5.4L Shelby 1000 at the New York Auto Show in April, along with the Shelby GTS and 50th Anniversary models. They followed up with 5.8L 2013 and 2014 Shelby 1000 and 1000S/C models along with GT500 Super Snakes. The product proliferation continued with a 'Wide Body' Super Snake. For the 2014 model year, Shelby American introduced the Shelby GT and GT/SC.

Nor was Ford standing still. The sixth generation Mustang made its worldwide debut on December 5, 2013. It finally featured the independent suspension that Carroll and I had wanted ten years earlier, and offered a 2.3L EcoBoost four-cylinder, a 3.7L V-6 and a 5.0L Coyote V-8. Production began in July of 2014 as a 2015 model. More importantly, the licensing deal Shelby and Farley had inked in 2011 resulted in the introduction of the Shelby GT350 at the 2014 LA Auto Show, and Shelby GT350R in 2015 at Detroit. Both models featured a fantastic 5.2L flat-plane crank, naturally-aspirated V-8 that put out 526hp. The GT350R model featured weight reduction through deletion of the radio, air-conditioning, rear seats, etc, and incorporation of carbon fiber wheels. Stickier tires, a larger front splitter, and rear wing improved downforce, and a limited-slip Torsen differential improved cornering. Only 37 GT350Rs were planned for the 2015 model year to match the build of the original 1965 GT350 race cars, the rest were built as 2016 models. The first GT350R was sold at Barrett-Jackson

Shelby 1000, Carroll's secret weapon. (Courtesy *Motor Trend*)

THE LAST SHELBY COBRA

for $1,000,000 with the proceeds going to the Juvenile Diabetes Research Foundation. I can promise you that Carroll would have loved driving the GT350R. He always appreciated light weight, and the responsiveness that went with it. We used to joke about enthusiasts whose primary motive was to brag about horsepower, knowing full well that on the street most of that power would literally go up in smoke. The same held true for Super Snake II: a naturally-aspirated Cobra would have been better balanced, but we agreed that if we wanted Super Snake II to get noticed, a supercharger and lots of horsepower would be the key!

2015 Ford Shelby GT350 and GT350R. (Courtesy Ford Motor Company)

Shelby GT350 with rear wing. (Courtesy Ford Motor Company)

Shelby American used the sixth generation Mustang to introduce a new Shelby GT, Shelby GT EcoBoost, and Shelby Super Snakes, and even a limited run of 50 2016 Shelby Terlingua models. Taking a page from Carroll's Chrysler days, a 700hp F-150 was introduced in 2015, followed by introduction of a Ford Shelby Raptor and Ford Shelby Baja Raptor. Product proliferation continues to this day, under the guidance of President Gary Patterson. Carroll's grandson, Aaron Shelby, also joined the board and is actively assisting with strategy, future planning and global expansion of Shelby American, as well as attending Team Shelby events.

Back in Detroit it was time to complete Super Shake II but the sense of urgency was gone. Post recession, business at Technosports had picked up and the shop was at capacity with high priority OE projects, including the build of the Ford Performance Technical Center in Concord, North Carolina. Technosports owner, Bob Nowakowski, graciously offered to continue with the completion of Super Snake II as fill in work at a very favorable rate. Over the next five years, we gradually completed the myriad details required to finish the car. We had a scare when two of the team members, and close friends, suffered heart attacks. Fortunately, they recovered quickly and thanks to modern medicine are better than new. The 5.4L Ford GT motor was upgraded with the Ford Racing performance package to reliably produce in the neighborhood of 700hp – that would get people's attention! Stainless steel headers were fabricated. Next we finished widening the body, bonded the honeycomb tub to it and mounted it to the Uni-Chassis with rubber isolators. Mechanisms

2016 Shelby Terlingua. (Courtesy Shelby American)

2017 Shelby Baja Raptor. (Courtesy Shelby American)

Super Snake II seats, console, and sub-woofer.

Super Snake II instrument panel.

were fabricated for mounting the doors, hood, and decklid, and side pipes were mounted. The battery was mounted in the rear, off the frame rail. For the interior, we wanted the car to look like no ordinary Cobra, and wanted to signal the marriage of classic Cobra with Ford GT technology. An instrument panel was fitted with the Ford GT gages, switchgear, console, and McIntosh radio. The carbon fiber Ford GT seats now fit in the widened interior, with Sparco 4-point harness attached to the honeycomb structure, along with a classic Cobra 427 roll bar. Aluminum door trim panels were fabricated to complete the interior design theme, with interior release latches inspired by those on the Daisy show car.

Then came the tedious part, connecting all the various systems. The complicated brake lines for the ABS, a center-mounted race fuel filler, coolant lines, fabrication of an air cleaner, and mounting the innumerable Ford GT electronic boxes. The windshield frame was widened and a tool made for the forming a new, wide windshield. Wiper patterns were established and new linkages fabricated. While we were at it, we designed and fabricated an adjustable anti-sway bar to be used for chassis development. Finally, aluminum panels were formed for the trunk and splash shields. By the end of the summer of 2017 Super Snake II was mechanically complete, and ready to be shipped for the installation of a custom wire harness that would make the complicated electronics work.

They say that if you want to make God laugh, tell him your plans for the future! Life certainly has its twists and turns, and as Super Snake II slowly progressed, other projects intervened in each passing year.

2012
Shamel Rushwin called and asked me to help fix an automotive lighting supplier that had been de-sourced by Ford and Chrysler due to its unscrupulous prior owner. We eventually got the company back into the good graces of Ford and Chrysler, and successfully sold the business to a competent new owner.

2013
Ford funded a mobility project with Lawrence Technological University, under the direction of Keith Nagara and myself. Called 'Quadricycle,' in deference to Henry Ford, we designed and patented a $500 folding electric vehicle for developing megacities like Lagos, Nigeria.

2014
In late 2013 I began working on a patent for an idea that had been in the back of my mind called 'Partial Forced Induction.' Computer simulations confirmed that turbo lag was reduced by 10-15%. I filed for a patent in 2014, which was recently published.

2014-2015
Walter Borda, my former attorney, came up with an idea for using automotive technology to develop a concussion assessment device, with the help of my friend, Tom Wallace, retired Corvette chief engineer, and Priya Prasad, renowned Ford safety technical fellow. Tom was relocating to North Carolina and asked me to take over design of the production concussions device. Correlation with evidence of concussion was excellent, but the wheels of sports and sports medicine turn slowly. Units are still on test gathering more data, and the military has expressed interest in using the device for assessing concussions on the battlefield.

I titled this Chapter 'Kismet,' which translates, from the Arabic, to fate or destiny. I don't believe in 'fate,' but, beginning in 2012, three serendipitous events occurred over the next few years. The first actually happened late in 2011 when an automobile manufacturer introduced a plug-in hybrid vehicle. I was astonished, and flattered, that it looked eerily like the hybrid chassis I envisioned and defined in my 2008 Uni-Chassis patent application. I doubt I told Carroll of my discovery at the time, as he was already ill. In 2012 my patent attorney sent a polite note to the manufacturer indicating that the vehicle might be infringing on my patent. The response coming back was slow and negative. Letters went back and forth over the next four years. It was classic 'David vs Goliath,' with the manufacturer waiting me out. Finally, having tried the gentlemanly approach, I took a page form Carroll's playbook and hired a respected Detroit law firm to take the case. Our first effort was to try and resolve the matter equitably as a win/win proposition – to no avail. Finally, as 2016 came to a close, we filed suit. The case is still pending in the courts. Hopefully, it will be resolved satisfactorily.

The second curious event was a little less dramatic. I had sent word to Carroll that I had found the molds for Daisy. After his death, I contacted the owners of the former Saleen facility to see if they were willing to sell the molds. They weren't interested, even though the molds had been left out back in a dump heap, upside-

Electric 'Quadricycle' built behind the historic Ford Piquette plant.

down, collecting rain and warping in the sun for ten years. They were probably useless, but I still wanted them. The guys at Technosports always questioned why we didn't use Daisy with its 10in longer wheelbase, rather than a classic Cobra body for the Uni-Chassis proof-of-concept. Fitting all the Ford GT components into a Cobra was like putting 10lb of 'stuff' in a 5lb bag.

Penske Vehicle Services acquired the Saleen facility in the summer of 2014 for final assembly of Dodge Hellcats and other specialty vehicles. Jill Lajdziak, a former GM Saturn Executive, was in charge of the acquisition and operation of the facility. On a whim, I called Jill and inquired about the Daisy molds. She knew nothing about them, but promised to find out. Some time later she called and said the molds had been disposed of, but they were getting ready to upgrade the facility to typical Penske perfection, and were willing to sell the model.

"What model?" I asked.

"There's a model in storage, and we need to get it out of the way for the facility renovation." Excited, I arranged to meet with Jill. We talked about business, the Ford GT built there, and the renovation plans for what was still a very nice facility. She offered to give me a tour of my old stomping grounds as we went to see the model. Finally, we came upon the full size rigid foam model of Daisy. It had been milled from Ford CAD data and used to cast the molds for the Daisy stunt props. It was dusty and a little worse for wear, but could be refinished to create new molds. After some negotiation, I acquired the model in February of 2015, and had Technosports store it at their facility. I hoped to someday build my own Daisy.

The third, and most unimaginable event occurred on August 29, 2017. We were just finishing up the mechanical build of Super Snake II to send to Dave

THE LAST SHELBY COBRA

Decker at Precision Race Service, for fitment of a new wiring harness. I spotted an article in the *ClassicCars.com Journal* by Larry Edsall, who'd written a book about the 2005 Ford GT. The headlines read, "V-10-powered Ford Shelby Cobra concept up for bidding in November." This couldn't be true. Ford had sold lesser push mobiles at auction before, even the fiberglass Shelby GR1 static model, but why would they sell the last Shelby Cobra, especially one that was a fully engineered driver? The article went on to say that "The 2004 Ford Shelby Cobra will cross the auction block in early November at GAA Classic Cars auction at the Palace in Greensboro, North Carolina." I'd never heard of the GAA auction. Why would they sell it there, instead of Monterey or Scottsdale? The article went on to say that "The winning bidder will write his or her check to the Henry Ford Estate, a 501(c)3 charity that will apply the money to the restoration of Fair Lane, the Dearborn, Michigan, estate of Henry and Clara Ford." The article also noted that, "Because the concept did not go into production, for liability purposes, Ford will 'render the car completely not drivable' before its delivery to its next owner, although the engine will remain functional." Now I understood the why, and my mind started to churn – maybe big collectors will miss this, or get scared off by Daisy not being drivable.

I decided to lie low, and see how much notice the auction received, never mentioning it to anyone. I dreamt that perhaps I might actually be able to bring Daisy home. Being a pack rat, I still had the Daisy and GR1 build books and prints. It would be a crime to have a one-of-a-kind, multi-million dollar concept car collect dust in a museum. Surely we could get it running again. I doubted I would be able to afford it, but perhaps I might get lucky, or perhaps Kismet was at work.

Over the next two months, I monitored web traffic. Daisy was the feature car for the GAA auction, but there had been little publicity. I asked my wife, Tracee, if she wanted to visit her sister, Betsy and brother-in-law, Jason Vines (formerly VP of Ford Public Relations), in North Carolina. On the way we could stop at the auction, and see what happens. At the last minute, I registered for the auction, hoping that the big guns might not show. We arrived at Greensboro the afternoon of November 2, and checked into The Palace

Opposite top Fake Daisy props, destroyed in the movie *xXx*.

Opposite bottom Rigid foam Daisy model at Penske.

Tracee standing guard!

auction house. Surprisingly, The Palace was a dedicated facility for classic car auctions, unlike the huge, temporary, tent affairs in Scottsdale and Monterey. In the showroom, hundreds of classics were on display, surrounding the permanent seating. The stage was up front, where the cars would be towed past the seated bidders. Daisy was front and center as we walked into the showroom. Acting like clueless Inspector Clouseau, my disguise was a plain black baseball cap, shirt and jeans, so I could surreptitiously scout out the crowd. My hopes were raised when I didn't spy any heavy hitters. The Shelby Cobra Concept, lot #FR0085, was scheduled to cross the block around noon the next day. We went back to the hotel anxious. Morning couldn't come soon enough, as I tossed and turned throughout the night, images of Daisy dancing in my head.

We were among the first to arrive at the auction the next morning, and strolled through the wide array of cars, many of which would have tempted me under normal circumstances. As the auction started, we wandered back toward Daisy. A commotion was going on around it: someone had taken the key! Ford's Greg Stoner, who had shepherded Daisy across the country, was crawling under the instrument panel searching for a spare key that might have been hidden. Failing to find one, they pushed Daisy to the far corner of the garage area, attached to the showroom. I wandered back to see what was happening, secretly hoping they wouldn't be able to start or hot wire Daisy. If the engine didn't fire up, maybe Daisy would be within my grasp at auction. My conscience kicked in, and as panic set in around the car, I introduced myself to Dean Green, owner of Greensboro Auto Auction and Greg Stoner (who had indeed recognized me). I told them I couldn't remember if Daisy was fitted with a transponder chip key, which might pose a problem if they didn't have the key code, as those keys required communication with the ECU to start. As a last resort, failing to find a way to hot wire Daisy, Mr Green had his technician from the fleet auction house come over, make a wax impression of the key cylinder, and cut a key to fit. It worked, and Daisy roared to life about an hour before it was set to cross the blocks. Although I hadn't helped, my conscience was clear.

Tracee and I took a seat near the center aisle behind the area reserved for dealers, and watched the classics roll by, taking note of the buyers. Finally, it was time for Daisy. As they towed Daisy in with Greg behind the wheel, the auctioneer started with a brief video of pictures from the 2004 press kit, followed by images

Daisy's proud new owners.

of the Ford Fair Lane mansion, closing with videos of Carroll driving the car at Irwindale. Greg raised the power hood to reveal the V-10 while Kathleen Mullins, President of the Henry Ford Estate, was introduced to emphasize that proceeds from the sale would go towards restoration of the mansion. I had promised myself that I wouldn't exceed a predetermined limit, and would stay calm. Instead, I got up, walked over, shook Greg's hand as he lowered the hood, and went over to one of the spotters and asked him to stay close to me. "How much are you willing to spend," he asked. "Probably not enough," I answered. Mr Green told the crowd that the winner would write a check directly to the Henry Ford Estate, it would be tax deductible, and there were no auction fees. Greg fired up the V-10 for the crowd and the auction began. The next five minutes were the most excruciating of my life!

Bidding ramped up quickly. I tried to hold my powder, but my first bid, at my predetermined maximum, came within the first 30 seconds. Bidding stopped for 15 seconds. I held my breath. Suddenly a couple of bidders behind me got active and the price started to climb. I managed to get one more bid in, but dejectedly walked back to my chair. The spotter followed me, as the price continued to rise, and then stalled.

"One of one … one opportunity," barked the auctioneer.

Tracee said, "If you miss this one, you're never going to get another chance."

The auctioneer stopped, and urged the crowd on, "You can't put a price on this … signed by Carroll Shelby, a piece of automotive history." Another bid came, and the price stalled again.

"I think you'll get it for another $25,000," cajoled the spotter.

"OK, this is crazy, one last shot," I replied. Another 30 seconds passed as the auctioneer raised the gavel, but barked on. The spotter and I kept motioning to drop the gavel. "I've got one man asking for the hammer and two begging for no!" yelled the auctioneer, hand raised again as they pushed Daisy off the stage. Finally, the gavel dropped, "Sold!" I jumped for joy. The crowd erupted with applause. The spotter and I hugged. It was the scariest, most irrational thing I have ever done.

People in the crowd came up to congratulate me, as the auctioneer announced that, "Moray Callum, VP of Ford design, will issue you a certificate of verification." I figured my old colleague would get a kick out of writing me that letter. For the next hour, my heart was still racing, alternately fueled by elation and fear of my snap decision. Tracee tried to calm me down. Kathleen Mullins came to thank me and offered a tour of the museum. A few reporters and the local video team sought me out for interviews. Eventually, I regained my senses. Two of the buyers representing a couple of the big collectors came over, introduced themselves and congratulated me. Had Daisy been running, they told me their buyers would have gone well over a million to get her. Greg Stoner congratulated me, provided some written operating instructions, handed over the keys, and kindly told me that Special Projects in Plymouth, Michigan had been instructed to disable the vehicle. I took the keys over to Reliable Transport and arranged for delivery to Technosports. Finally, it was time to pay the bill. I maxed out the home equity loans on both our house and cottage, and cut the two largest checks I'd ever written, to the Henry Ford Estate.

After the adrenalin rush wore off, it was time to head off to see Betsy and Jason in Wilmington. On the way, we made calls to our boys, plus Betsy and Jason to tell them the news. I also called Bob Nowakowski to tell him Daisy would be arriving on Monday. The final call was to Ken Yanez, owner of Special Projects. I'd known Ken for over thirty years, having worked with him on my first concept car, a DeTomaso sports car at Chrysler, but hadn't talked to him in quite a while. Although I had a pretty good idea of what had been done to Daisy, I told Ken about my wild acquisition and asked how they disabled Daisy.

"I don't know, but I'll ask the guys," he responded. I knew he would, and knew that Daisy would run again. That night we celebrated with Betsy and Jason, and finally got a good night's sleep.

Back in Detroit, I couldn't wait to get to Technosports. When I arrived, Mike Nowakowski (whom I affectionately called 'Mikey') already had Daisy up on the lift. Mikey had assembled the running chassis 15 years earlier, and remembered all the details. We noted that the service plate on the torque tube had the bolts ground off, welded over, and ground flush. It took less than an hour for Mikey to get the service plate off to inspect the driveline. As expected, the coupler between the output shaft and driveshaft had been slid back, but surprisingly welded to the shaft spline. The output shaft spline had also been welded and was ruined. Getting the torque tube out was a more difficult proposition, as the rear transaxle had to be removed, which necessitated disassembling most of the trunk area with hydraulic pump for the hood and AC/DC show car electronics. Days later, the torque tube was out, revealing the signatures of Manfred Rumpel and the rest of the Daisy chassis team.

As this was going on, word of my acquisition of Daisy began to spread. The sale had been picked up by the press, and many of my Ford colleagues called to congratulate me. Other Daisy team members offered to help get her back on the road. Edsel Ford sent a kind note, "Dear Chris, How exciting that you were the winning bidder for the 2004 Shelby Cobra Concept car. Congratulations! More importantly, your gift has made a huge difference to the capital campaign at Fair Lane and it has taken us almost to the end of our Phase 1. I don't know where you are living nowadays, but if you are around, I would be pleased to show you around Fair Lane. The house looks fantastic and I think my great grandparents would be very pleased." I thanked Edsel, and we agreed to tour Fair Lane in the spring. John Clinard called from the west coast PR office, "Congratulations on Daisy! So good that you

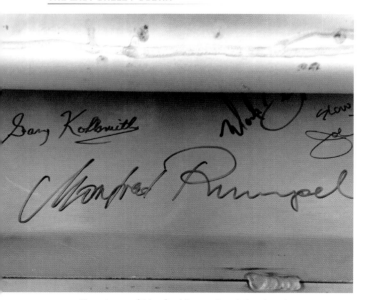

Signatures of Manfred Rumpel and the chassis team.

Carroll's autograph from 2003.

have that car!" Moray Callum called to say he would be sending the certificate of authenticity. "If I knew you were going to be the buyer, we wouldn't have buggered the driveline so badly. The lawyers wanted us to put cement in the block, but fortunately we talked them out of it!" I called Bill Warner, who knows almost everything in the collector car world, to tell him the news. "I didn't know Daisy was up for sale," he said with surprise, "if you get her running I'll make a spot for her at the Concours!" Now we had a deadline, and had to move fast. I suspended all work on Super Snake II so we could focus on getting Daisy running for the Amelia Island Concours d'Elegance.

While Daisy was in great shape aesthetically, mechanically there was much to be done after 14 years of show appearances. The clutch, of course, was fried (probably by Carroll) and the clutch fluid had turned to jello. The hydraulic pump for the power hood had leaked over the transaxle below, shredded rubber coated the rear of the chassis, and surprisingly, there were small rust holes on the top of the vertically oriented mufflers! The long-lead item for making Daisy drivable was obtaining a new output shaft from Livernois Engineering, supplier of the original. We provided the prints and the destroyed shaft, but racing season was about to begin, and they were backed up with work. While we waited over the holidays, Mikey installed a new clutch and flywheel, new plugs for the V-10, replaced all fluids, installed new mufflers, and cleaned her up. Livernois delivered the output shaft in late January, and we began putting Daisy back together.

By the end of February, Daisy was complete and ready for a test drive down Autry Street in front of Technosports. It was great to get back behind the wheel, and I must admit that I lit up the tires once or

Transaxle back in Daisy, and all cleaned up.

KISMET: DAISY COMES HOME

New plugs and fluids, ready to roar!

twice going down the block. Fortunately, the car was stuck in 'valet' mode, so I couldn't get too crazy. We'd take care of that after the show. The reliable carrier arrived March 6 to take Daisy to Amelia Island.

The forecast for Sunday was rain, so, as a precaution, the concours was pulled ahead to Saturday. Bill Warner had kindly arranged a prime display spot inside the entrance of the Concours. We unloaded Daisy on Friday afternoon, and I drove it past the crowd at the Ritz-Carlton to its appointed spot on the field. The crowd was already pointing and cheering as I passed by, revving the V-10. I spent the rest of the afternoon and most of the evening answering questions from admirers, with no time for a pit stop. We covered the car for the night (except for a 2am beauty shot Bill had arranged), praying for a dry evening. The next morning crowds started to gather around Daisy before the show even opened, and the traffic continued throughout the day. Admirers seemed to fall into two categories: those that remembered Daisy and recalled the *Rides* show, and those that were seeing Daisy for the first time: "Is it coming out next year? ... How come they didn't build it? ... How much? ... Where can I get one?" they asked. Many of my friends stopped by to see Daisy: Barry Meguiar, Beau Boeckmann, Luigi Chinetti, Moray Callum, Ralph Gilles, and others – including Mark Reuss, President of GM. It was a long and exhilarating day talking to all the Shelby admirers. Barry Meguiar invited me to the Gala dinner where Emerson Fittipaldi was honored. Even more significant was the standing ovation the crowd gave to Bill Warner, founder of the concours, who, exhausted from the event and his successful battle with cancer, was moved by the showing of love and appreciation.

Clearly Daisy had aged well. What some people had thought was too contemporary 14 years earlier, now looked modern and desirable. It was the response and questions of these enthusiasts that inspired me

Test driving Miss Daisy! (Courtesy Howard Walker)

Mikey, Gene Duprey, the author, and Brian Hermann.

to finish this book. Shelby enthusiasts are interested in Carroll's workings over the last quarter-century of his life. Daisy was invited to the July 2018 Concours d'Elegance of America in Plymouth, Michigan. The crowd response reaffirmed the ongoing interest in all things Shelby.

Daisy is now home. We fitted a new wiring harness to eliminate some non-automotive wiring the Valencia studio installed in the rush to get Daisy ready for its debut in 2004. In the process, we found a keyed control box switch labeled: V, L, and U … Valet, Limited (100mph) and Unlimited! You can guess what we selected. The show car wheels and tires were switched to the development wheels and tires, like those used during Carroll's Irwindale drive. I can't wait for some warm, dry Michigan weather to let this snake loose!

Writing this book brings back a lot of memories. I cannot read a Shelby quote without my inner voice reading it in his East Texas drawl. The stories conjure up his big smile, and the jokes come with images of those jiggling shoulders and impish giggle. Google Carroll Shelby, and the first thing that pops up is "American Automotive Designer," which Carroll definitely was not. I once asked Carroll how they designed the flares on the Cobra. "We didn't design nothin'," he said, "the chassis was so flexible that we just kept putting on bigger and bigger tires. The guys just banged out the

Opposite top Beau Boeckmann and Chris Theodore at Amelia.

Opposite bottom Barry Meguiar checking out Daisy.

Daisy at Concours d'Elegance of America. (Courtesy Don Wood)

fenders to fit." Truth be told, whether in the 1960s or 2000s, Shelby wasn't terribly interested in styling, and left that up to the designers (Peter Brock and Vince Laviolette in-house, plus Ford and Chrysler designers). He expressed his opinions, of course, but was focused on performance. Nor was Shelby an engineer. He learned by doing – seeing what worked and what didn't.

So who was Carroll? First and foremost, he was a winner and a fighter. He only played to win, whether it was driving a race car, managing a team, building performance cars or starting a business. A serial entrepreneur, he was willing to take big risks. If he believed in something, he would go 'all in.' On the wall over his desk in Gardena was the sign: "Will those who say it can't be done, please get out of the way of those of us who are doing it." There have been other great drivers, team owner, car builders, and businessmen (Ferrari, Chapman, Penske, and others come to mind), but I can't think of any that seemed to enjoy every minute – whether winning or losing, happy or sad – as much as Shelby did. Carroll just loved the adventure of life. He was having fun!

Second, Carroll was a leader. The goal was always to win, and Shelby pushed his people. He was great at recognizing talented drivers, and, through luck or good judgement, knew how to put together great teams. He delegated and expected results, never getting into the details or micro-managing. Perhaps Shelby's good fortune in picking up the remnants of the Reventlow race team made him a little too trusting of others in some of his later dealings, such as with the Series 1, but Carroll learned from his mistakes. Even Shelby's failures somehow became cannon fodder for great stories.

Finally, Carroll was a people person. A great salesman, marketer, communicator, and a unique, unforgettable personality. His circle of friends was huge, and he kept in constant contact with all of them. He had a genuine appreciation for people in every walk of life, and it seemed like everyone could relate to him. Even the few who hated or felt somehow wronged by Carroll, begrudgingly admired him and his personality.

Reliving my time with Carroll feels like reading a tall tale from a story book. Thank you Carroll for your friendship and mentorship. Thank you for all your humor and unforgettable stories. Thank you for the opportunity to work with you on Daisy, and all the other projects. Thank you for encouraging me to patent Uni-Chassis. With a little luck, or perhaps good old Shelby magic, Uni-Chassis will help pay for bringing Daisy home!

Carroll, love ya!
Chris

KISMET: DAISY COMES HOME

Carroll with his original Cobra ...
... and the last Shelby Cobra. (Both courtesy *Motor Trend*)

EPILOGUE

If there is one thing I learned from Carroll, it is to keep looking forward. He had stolen a line from Frank Lloyd Wright, stating his favorite car was always, "the next one." Shelby always enjoyed juggling things, keeping all the balls in the air at one time. I believe this was one of the keys to his longevity. Between his projects, interests, and huge circle of friends, he was always busy, never bored.

I, too, am preoccupied with the future. It starts with a bucket list of projects I want to do, or complete, but includes adventures with family and friends that have yet to be experienced. Finally, I cannot help ponder the future for my passion (automobiles) and our society. The list is far too long, so I will prioritize along the way, but know that I will never run out of things to do.

Finishing Super Snake II, of course, is at the top of my list. Time to get her running, develop the chassis, make her pretty (or pretty mean) and take to the road. Next on my list would be the Shelby GT, so we can show enthusiasts what might have been if we had done a continuation of the Ford GT with Carroll. The GT had great bones, and I am convinced that with the updates that had been planned, the Shelby GT Mk IV would compare favorably with today's homogenous, high-tech supercars. The Shelby GR1 needs to be on the road too. It is, arguably, the best-looking car I ever worked on, and I believe it outshines contemporary grand touring cars. Finally, I still believe the world of hot rod and custom car enthusiasts deserve a modern aftermarket chassis to help them realize their design dreams. There are, of course, future car designs in my head that will probably never see the light of day, but perhaps with today's 3D printers, these ideas can take on physical form.

Nor is my bucket list limited to automobiles. I told Dan Gurney that I had an idea to take 'Gator' to the next level, but never had the chance to share the details. There are half a dozen other patentable ideas still on the to do list, but changes in intellectual property laws have made this more expensive and less rewarding – better to just encourage others to put their ideas into practice.

Finally, I believe we need to focus technology on developing countries. These nations are being left behind in a world focused on electronic aids for those that already enjoy a high standard of living, leaving over half the world's population behind. This approach is not sustainable. I believe we can use technology to provide a low-cost means for people in underdeveloped countries to meet the most basic of Maslow's needs, so that they can create their own new economy to supply goods and services to an ever-growing population. An ultra-low cost vehicle is only a minor part of the solution. While I cannot be directly involved, perhaps I can help others come up with a solution. An experiment in Shelby's beloved Africa might be a great place to start.

With the advent of autonomous technology, people often ask about the future of the automobile. In particular, enthusiasts want to know what will happen to performance cars, like those Shelby developed. I tell them there is good news and bad news. The good news is that autonomous cars have been overhyped, and will not become ubiquitous nearly as quickly as advertised. There will be progress, but unforeseen setbacks along the way. The bad news, which I have been forecasting for quite some time, is that, by 2050, driving on public roads will be banned in the name of safety – as it should be. In the hands of flawed humans, vehicles are lethal weapons. Once autonomous vehicles prove safer than human operators, the only thing inhibiting a driving ban will be the time it takes to replace the entire fleet of vehicles with autonomous cars.

So what does this mean for road cars and performance cars in the future? In an autonomous future, all road cars will accelerate, brake and corner at the same speeds, so performance will become irrelevant. There will be a combination of shared

EPILOGUE

vehicles, autonomous taxis, and privately owned autonomous cars. Products will be differentiated on aesthetics, comfort, convenience, features, and price – think of cell phones as an example. Eventually, vehicle architecture will change. Without accidents, there is no need for bumpers, crush structures, air bags, and perhaps even seatbelts. The vehicle footprint will be optimized to maximize passenger and luggage space. Envision different plastic modules mounted on electric skateboards. Enthusiasts may fear this future, but many will come to appreciate its convenience and other advantages.

So where does this leave performance enthusiasts and companies like Shelby American? The answer is 'car farms' like the horse farms that preceded it in the last century. Car condominiums with dedicated racetracks are already dotting the countryside. Enthusiasts will be able to have a place to store their prized possessions and a place to enjoy driving them – perhaps even a few 'private' roads will survive. Like horse farms, however, car collecting will become a more expensive hobby. There will still be high-performance cars built and sold for enthusiasts. They will fall into two categories: autonomous or semi-autonomous supercars that will become little more than amusement park thrill rides for the wealthy, and bare bone analog vehicles, like the Shelby Cobras and cars of its ilk, that demand driver skill to extract the most from their performance capabilities.

Automobile racing will change, as well. In my mind, racing has been in decline since the 1960s. Technological innovation has largely been banned. Race cars are unrecognizable advertising billboards. Races have become parades, with little passing and even less spectator excitement. Budgets largely determine which manufacturer will win, and the physical limitations of drivers are being approached with increasing g-loads. I suspect a new format will emerge – one where vehicles have less performance and are more difficult to control, requiring driver skill to make the difference between winning and losing.

For the enthusiasts reading this book, I believe we have seen the golden age of performance cars and automobile racing. New sports will emerge for future generations. In the meantime, the rest of us can enjoy and appreciate the times when men like Carroll Shelby challenged themselves to drive faster, innovate, build better cars, win, and have fun doing it.

APPENDICES

Appendix I

Transcript of Roy Lunn Letter

April 2004
The Original Ford GT 40
Roy Lunn

There have been numerous articles and books written about the period 1963 to 1969 when the original Ford GT was created, raced and became part of history. Since that time, I've been interviewed many times about exactly how it all got started. Now, a new wave of inquiries has materialized with the advent of the new Ford GT sports car and I find myself again providing the same material which is sometimes difficult to recognise after someone else's interpretation. I therefore decided to organize the facts before some of the distortions became reality. Doing this will also save me work should there be further enquiries. So, to mark the fortieth anniversary of the Original Ford GT I am setting down the following brief history from the notes I made at that time and extracts from SAE paper 670065 which I wrote in 1967.

40 years ago, in the spring of 1964, the first Ford GT 40 prototype was completed. It had been incubating since the spring of 1963 when a chain of events set the project in motion. It was the time when a vibrant new force under the leadership of Lee Iacocca was starting to change the complexion of Ford Motor Company. His main lieutenant was Don Frey, a young skilled administrator who had succeeded Lee as the head of Ford Division. He, in turn, had an energetic head of Product Planning working for him. His name was Don Petersen who later became the Ford President. These three young Executives were combining efforts to rejuvenate the company to produce exciting new vehicles with product leadership. The Mustang was the most noted of these and, naturally, all corporate creative activities were being geared to generate other innovative products.

I was fortunate to have been the engineer in charge of Ford Advanced Vehicle Development at this point in the Company's history. My department had then recently been responsible for a variety of exciting developments including the Mustang I, Big-Red – the futuristic superhighway truck, and the Cardinal which was Ford's first small car employing front-wheel drive.

The exciting forces at work at Ford at that time recognized the importance of racing both as a product development tool and a promotional medium. Although hampered by the then existing antiquated industry ban on racing they were already looking at ways of how to get back into Stock and Drag, and they were naturally looking at the worlds racing scene.... particularly at the long distance endurance events of GT racing. These events had long been the field of European manufacturers and at that time were dominated by Ferrari who was a consistent winner at famous events such as the 24 Hours of Le Mans.

It was in early 1963 that discussions first began to get serious. I was called in as head of Ford Advanced Vehicles engineering and also because I had come from Europe and had experience of designing GT racing cars among these, the DB2 Aston Martin and Jewett's RI Jupiter.

The desire to quickly get involved led to the possibility of buying the existing best in the field, namely Ferrari. A small team consisting of Don Petersen, Bob McGuire from Styling and an inevitable Finance representative and myself were hurriedly shipped off to Italy on the 12 April. Phil Paradise, the resident Ford manager in Italy made the arrangements and became a valuable member of the team. Lee Iacocca jokingly warned us that the atmosphere there would be hostile as it would be if an Italian Mafia group came to the US to buy the New York Yankees. The team visited several Styling Houses, Maserati and the main target of Ferrari in Maranello. Apparently, Enzo Ferrari had been making noises that, as he had no surviving heirs, he would like to rid himself of running the production factory and concentrate on the racing which was where he had started. Two days of reviewing the Ferrari facilities and having initial negotiations, was an interesting experience. It was also interesting to learn that Italian corporations seemingly have three sets of books. One they show the IRS. Another they show the potential buyer and lastly, the real set.

On returning to Detroit, the team reported that Ferrari had an interest in a combined Ford-Ferrari deal. Don Frey made a quick trip to Maranello and also brought back positive reactions. This resulted in assembling a team of mostly legal and financial people to go to Italy to develop and consolidate an agreement, but the expedition failed. Maybe because Ferrari got teed off with the nitty gritty of the lawyers and financiers or maybe he finally could not bring himself to relinquish control. Regrouping in Detroit the question was what do we do now?

I advanced the argument that if we had bought Ferrari and he had gone on winning, the success would've been attributed to Ferrari. If, however, the cars started losing it certainly would've been because we, Ford, had bought them. Also, as a confident engineer I stated that we had the capability of doing it ourselves in which case if we won it would be rightly Ford Corporation's achievement. If we lost, we would get the deserved blame. After debate, it was agreed to proceed in this direction and I was charged to draw up a proposal.

The basis for the proposal was to start a small group that could effectively draw on selective talents of individuals throughout the Company. We could take advantage of the financial support of the main Corporation while operating with the freedom and agility of a small independent company. The eventual aim was to have a vehicle created by Ford and constructed in Dearborn. It was emphasized however, that initially this group would have to operate in Europe and tie in with an existing company having GT racing experience and direct supplier industry ties. This last factor spawned a trip to Europe to review possible candidates. On 13 June, Ray Geddes the business manager for Special vehicles, Hal Sperlich manager of product planning and I departed on this mission.

Colin Chapman of Lotus was first on our list because Ford already had ties in this direction with the Indy Car program. However, Lotus did not have direct GT car experience at that time, plus the fact Colin was very independent and not easily converted to a team player. Cooper was reviewed and although they were conducive to such an arrangement, they did not have direct GT car experience. Lola was the third visitation and the main objective of our search. Lola was owned and run by a bright, innovative designer named Eric Broadley who had produced a mid-engine GT vehicle of the variety we desired, but unfortunately had limited facilities. Recognizing that we, Ford, could easily rectify this deficiency Lola became the partner of choice.

We returned to Dearborn and Don Frey mad the presentation to Lee Iacocca's Executive committee on the 12 July. The plan to proceed on our own was presented with the objective of having a car running in one year. I was amused how quickly the plan and the budget I had prepared was approved … literally a matter of minutes. The conversation then centered on the PR aspects and lasted another 40 minutes or so. By the end of that time the objective of having "a running car in one year" had been stretched to "racing in one year" and with grand optimism of winning in that time period.

Don Frey, Ray Geddes and I left on 16 July on one more trip to England to finalize the arrangements.

While all these events were taking place, I had a group in the Advanced Vehicle department working on the program and they were developing the packaging of a midship engine vehicle layout derived from the Mustang I. Also, Gene Bordinat VP of Styling had Johnny Najjar and his team working on full sized clay based on the hard points we fed them. These guys quickly picked up the necessity of "form follows function" and by the time we tied in with Lola they had a full-sized model ready to ship. Also during this period Ford ties with Carroll Shelby had developed with his Ford engined Cobra cars which incidentally, were also designed and produced by AC cars another English source. Carroll had direct GT racing experience by driving an Aston Martin to victory at Le Mans. He also had direct ties to John Wyer who was the Aston Martin racing manager when Carroll was driving. It was this chain of connections that led us to sign up John Wyer as the Resident Manager of Special Vehicles in England when we set up shop there.

John's first assignment was to find an expanded facility for moving Lola into as well as a new facility for Ford Special Vehicles operations. He fortunately found a new small factory duplex in Slough on the west side of London airport. This was convenient for the International travel that would be involved, but its main advantage was its location near the main roads to Birmingham and Coventry where most of the parts suppliers are located. Setting up the Slough facility, moving Lola into half and the Ford activities into the other was all accomplished in a remarkably short time. I moved to England with the Ford Advanced Vehicle engineering team, which included my right hand man Chuck Mountain, and industrious young engineer, Len Bailey an excellent designer who was originally from England. After his work on the Ford GT Len decided to stay in England and became a well know designer and maker of specialty cars. We brought in other designers and individual specialists from various Ford divisions on an as needed basis.

The contract with Eric Broadley was for him to work exclusively for Ford for one year. In addition, he was to provide the two prototype midship engine cars he had built the year before to be used as development test bed for the Ford components. In return, he was to be moved into a new, upgraded facility in addition to receiving appropriate remuneration. After the Ford program, Eric went on to become the world's leading supplier of Indianapolis 500 cars and a broad range of other racing vehicles.

One of the Lola prototypes, which had run briefly at Le Mans the previous year, was adapted with a new Ford power plant and briefly tested before winter set in. Meanwhile, the design and build of the first Ford GT 40 continued. Some systems were shipped from the USA while others came from local suppliers in Europe. For example, the transaxles were made by Collotti in Italy and were the same as used on the Lola together with the drive shafts couplings made by Metalastic in England and the spoked wheels by Bararni of Italy. We made this choice because we had problems with the original cast wheels. The body skins were made by a local supplier who was a very capable specialist in high strength, low weight fiberglass.

Back in Dearborn, the racing embargo had been ignored or rebutted and Ford had formed in Special Vehicles Department to cover all aspects of racing. It included stock cars, drag, Indy, rallies and now GT racing. While the Ford GT was being developed, the Shelby Cobra was filling the gap. These cars were doing well in the production category, but were no match for the outright prototype class dominated by Ferrari.

Carroll Shelby naturally became the principal choice to eventually race the Ford GT 40s (as they were now called) and became directly involved by having Phil Remington join our group in England. Phil was not only an extremely nice guy to work with, but was a very talented racing engineer with a wealth of experience. He became a major contributor to the program and the very important link when the cars emerged from the build shop and went to the racing team.

In the early testing and development work, Bruce McLaren was the test driver. He provided excellent inputs and became the continuity into the driver's seat for racing.

The first GT 40 was completed on 1 April 64, some eleven months after putting pencil to paper in Dearborn and we were all looking forward to being able to start development driving and testing. But this was not to be. I had received the direction that it was to be first prepared and flown to New York for an announcement introduction. Of course, this stole valuable time from the program for going to the Le Mans test on 16 April. So, the car was polished and made to look pretty and we all boarded a plane for New York.

Lee Iacocca made the introduction announcement and the car was extremely well received. But, as I sat there in the audience and listened to the glowing words, the fact that it had not yet turned a wheel worried me.

By the time we returned to England there was hardly any opportunity for proving the vehicle before packing it off to France for the test weekend. The second car was completed on 11 April and it, too, was hurriedly shipped to Le Mans. The first day of practice. 16 April, proved to be rain drenched and after very few laps, the first car totally wrecked on the Mulsanne straight when it left the road at over 150 miles per hour. The second car also experienced trouble but suffered only minor damage. Luckily, both drivers were unharmed, but obviously, some stability phenomenon existed that was not apparent during the design analytical phase. The problem and the solution were found within one week of returning to England, where further testing was carried out at the MIRA proving ground. The fault was found to have been an aero stability condition, which caused a rotary motion of the rear end of the vehicle comparable to that of an arrow without feathers. This motion increased with speed and, accentuated by the wet track, eventually resulted in rear end break away. Subsequently, it was found that the adaptation of a rear end 'spoiler' not only had the effect of putting feathers on an arrow, but also slightly reduced drag. Apparently, the 'spoiler' created and air tail which artificially increased the vehicle's aspect ratio and moved the center of pressure rearwards. It also increases the adhesion of the rear wheels and, surprisingly, the effect of this small addition could be felt down to 70mph.

The remaining car from the Le Mans practice was modified by the addition of the "spoiler" and rebuilt in readiness for the GT 40's first race outing at Nürburgring on May 31, 1964. The car performed most favorably in practice and qualified second only to the fastest Ferrari. It also ran second in this 1000-Kilometer race in the early hours, but retired after 2 ½ hours. A suspension bracket failed because of an incorrect welding process, but when the vehicle was examined there were several other areas showing distress and near failure. The outing was, therefore, most successful as a development exercise, and the lessons learned were quickly incorporated into the three vehicles being built for the Le Mans race in mid-June 1964. These vehicles were completed and weighed in at Le Mans scrutineering at 1960lb, less driver and fuel.

In practice, the cars qualified second, fourth and ninth. During the race, one car held the lead for the early hours before retiring with a transmission failure. The second car retired after five hours with a broken fuel line and the third car retired after 13 ½ hours with transaxle problems but not before establishing an all-time lap record.

Every attempt was made to correct the transaxle problems within the limited time available before the next race at Rheims, France, on July 5, 1964. Again, the cars led the race in the early hours and set new lap records but all retired with transaxle failures. In addition, the nature of this circuit showed insufficient cooling of the brake discs which remained red hot during the entire time the cars were running.

The GT 40's first season of racing 1964, therefore, showed seven starts in major events with no finishes. The cars had demonstrated that they met the performance objectives, but failed badly on durability aspects. The winter of 1964 was devoted to detail preparation of the cars for the 1965 season and at this stage, the responsibility for racing the vehicles was given to the Shelby-American racing team. Twenty-one modifications were executed on the transaxles; the rubber driveshafts were replaced with Dana couplings, and the decision was made to install standard 289 CID. cast iron engines, as used on the Cobra, using wet sump lubrication. New cast wheels were also installed and increased to 8-inch front and 9½-inch rear rims. Two of these cars made their first appearance in the 1965 season at the Daytona 2000-Kilometer Race on February 28, 1965. They finished first and third in this event, setting an average speed record of 99.9 mph for the distance in 12 hours and 20 minutes. Two vehicles were also entered in the Sebring Race in March 1965 and finished second over-all and first in class, once more demonstrating that a fair degree of durability had been attained. The cars were raced by the Company once more in 1965 at Le Mans, but without success.

The decision was then made to manufacture 50 of these cars in order to qualify them for production sports car category. The cars were completed in the 1965 period after detailed changes and the adoption of the 5-speed ZF transaxle. These GT 40s were sold to the public, and, in the hands of private race teams and individuals, won the World Championship for production sports cars in 1966.

In the fall of 1964, the engineering team relocated to Dearborn and started operations as Kar-Kraft, a Ford contracted facility. This team continued engineering on the GT 40 and also started working on a new experimental vehicle.

The 1964 season had shown the GT 40s were currently competitive on performance factors, but lacked durability. Although work was progressing on correcting the durability problems, it was obvious that the GT 40's performance in the fast-moving racing field would soon be outmoded. The problem was, therefore, how to get an improved power-to-weight factor and at the same time achieve a high durability level. The alternatives were to generate more power from the 289 CID series engine or adapt the 427 CID engine which had been developed for stack car racing. This latter approach would also involve the development of a unique transaxle to handle the higher power. The other indeterminates were whether the additional weight (some 250lb) for the larger engine and heavier transaxle and driveline would unduly deteriorate handling and accentuate braking problems. It was decided, however, to explore this approach by constructing a test vehicle and physically evaluating its performance. The program was initiated in the winter of 1964 and was designated the Mk II project. At the outset, it should be emphasized that the exercise was intended to generate information for a future model, and there was no intention of racing the car.

Package studies showed the 427 CID engine could be accommodated in the GT 40 basic structure by modifying the seating position and rear bulkhead members. The basic suspension units were unchanged, but provision was made for the 8 inch wide cast magnesium front wheels and 9 ½ inch rear wheels. Housing of the wider spare wheel necessitated revising its position, and the new front-end arrangement mad provision for a remote engine oil tank of the bulkhead and a larger radiator.

A major problem was to generate a transaxle unit which would handle the 427 CID. driveline was used, but with completely new housings and axle unit. Ed Hull, our head designer, did a great job on the design development of the new transaxle based on the stock car transmission unit with its 4 speed synchronized ingredients. This approach resulted in a heavier and less efficient arrangement than a direct transfer box, but had the advantage of using already developed components. The housings were designed in magnesium, and a pair of quick-change gears transmitted the power to the pinion shaft. The resulting over-all package from these changes required new front and rear structures and bodyshell.

The first experimental Mk II vehicle was completed during April 1965, and was evaluated on the 5-mile oval at Ford's Michigan Proving Ground. After only a few hours of tailoring, the car lapped the circuit at an average speed of 201 ½ mph and exceed 210 mph on the straightaway. Subsequent testing on road circuits showed that handling had deteriorated only slightly. From these test results, it was calculated that this vehicle should be capable of lapping the Le Mans circuit in 3 min. 30 sec. to 3 min. 35 sec. without exceeding 6200 rpm. It these lap times could be realized at this relatively low engine rpm, the car would obviously have a high potential to win at Le Mans. The decision was made, therefore, to attempt to run two of these experimental cars in the 1965 Le Mans event as exploratory exercise. Since this decision was made at the end of April, the cars would be going to the event without even the benefit of the Practice Weekend.

In the remaining five weeks, the first car underwent initial testing and rebuild, and a second car was hurriedly constructed. The second car actually arrived at Le Mans without even having turned a wheel. Having missed the April practice, the first evening of pre-race practice was spent in tailoring the cars to the circuit and applying side and rear end fins for aerodynamic refinement. On the second evening, the car that had never turned a wheel before arriving at the track set and all-time record lap of 3mins. 33 sec. – an average of 141mph.

One car qualified first, and when the race started on Saturday, both cars went out ahead of the field and comfortably lapped at 3 min. 40 sec, without exceeding 6000 rpm. Unfortunately, hurried preparation resulted in the cars being retired after two and seven hours, respectively, with non-fundamental driveline problems. One car had a spec of sand in the clutch slave cylinder which caused the piston to stick and generate heat at the throw-out bearing. The heat in turn softened an oil-retaining ring in the axle ultimately resulting in loss of oil. The second car broke a gear which had been incorrectly drilled. The cars, however achieved their purpose of establishing the capability of the engine-driveline combination. The potential indicated in this initial experimental outing resulted in the 1966 program being based on Mk II vehicles.

The following chart shows the Mk II power-to-weight factor compared to the original GT 40 and the production version. Vehicle weights are quoted, less fuel and driver.

	Vehicle Weight	HP	HP/lb
Mk II	2400lb	485	.202
Production GT 40	2150lb	385	.179
Original GT 40 Prototype	1960lb	350	.179

In preparation for 1966, a concentrated vehicle development program was planned using the Daytona, Sebring and Riverside tracks. In addition, specialized component developments were initiated on items such as engine (SAE Papers No. 670066, 670067) ignition and electrical system (SAE paper No. 670068), transaxle (SAE Paper No. 67069) driveshafts and brakes (SAE Paper No. 670070). Although some fundamental changes emerged from this development program, the main emphasis was on refinement to establish durability rather than improve performance.

A major contribution to speeding up development was originated by the Ford Engine and Transmission engineers. They evolved a dynamometer, which could run the engine and driveline units under simulated road conditions that had been recorded on tape in an instrumented vehicle (SAE Paper No. 670071). This device allowed component testing to proceed independently of vehicle availability and climatic conditions.

Major changes that resulted from testing and

development executed by the racing teams at Shelby and Holman and Moody, included:

- New, shorter nose configuration to save weight and improve aerodynamics
- Addition of external rear brake scoops
- Higher efficiency radiators
- Strengthened chassis brackets for durability
- Live rear hubs for improved durability
- Internal scavenge pump to minimize vulnerability and save weight
- Generally improved ducting to radiators, carburetors and brakes
- Crossover fuel system with a single filler neck,
- Ventilated brake discs to improve durability
- Quick-change brake disc design to facilitate changes during pit stops

All of these changes were incorporated in the vehicles that made their first appearance at the Daytona 24-Hour Race on February 5-6, 1966. The Mk II cars virtually led the race all the way, finishing first, second and third for their first victory. It is interesting to note that in 1965 when this race was approximately at the halfway point, the winning GT 40 averaged 99.9mph. In 1966, the event was of 24 hours duration and the winning average was 109mph. This indicates the fast-moving nature of this field of competition.

The second race in 1966 was the Sebring event, where Mk II's finished first and second, setting new distance and lap records; a GT 40 finished third overall. The car that finished first was an open version of the Mk II with an aluminum underbody that was designated the XI.

One car was entered in Spa 1000-Kilometer Race and finished second.

After attending the practice session in April, eight cars were prepared for the Le Mans event that took place on June 18-19, 1966. Mk II's qualified in the first four places and set a new lap record 3min. 31 sec. or 142mph. The race took place in cloudy weather with intermittent showers during the 24 hours. Mk II's finished first, second, and third and despite the weather conditions, established a new record for the 24 hours of 126mph (previous best was 122mph on a dry track).

As a result of winning Daytona, Sebring, Le Mans and finishing at Spa, the Mk II's also won the World Championship for prototype cars in 1966, thereby meeting the original objectives set forth in 1963. The eventual aim, however, was to have a vehicle totally created by Ford and constructed in Dearborn. This final piece of the objective equation would necessitate developing an all-new vehicle which could take advantage of the revised FIA (Federation International Automobile) rules. This could contribute to the much needed weight saving and provide improvements in aerodynamics.

The mechanical components as developed for the Mk II were essentially all American creations and US based. The body structure was however still modified from the original GT 40 steel monocoque and still made in England.

My long-term associate Chuck Mountain had been researching materials for a new body structure and found an interesting possibility with aluminum honey-comb produced by Brunswick, Corporation who had developed it for aircraft applications. We studied Chuck's proposal and decided it was in the final major piece of the design jigsaw puzzle we needed for the new vehicle.

This material necessitated us to change our whole design approach because instead of welding, riveting and bolting the pieces together, they were glued. This became particularly tricky for highly stressed mounting points for the mechanical hardware such as the engine, driveline and suspension.

The new vehicle was originally named the "J" Car, but acquired several names in its evolution. For a period, it was called the Bread-Van before finally emerging as the Mk IV. These stages all used the same basic vehicle chassis, but with variations of the detachable, front and rear body skins.

The Mk IV program started at the same time as the Mk II's and ran parallel to it. As previously stated, the Mk II was only intended as a development test bed for the next generation of mechanical units. After a disappointing showing of the Mk II's at Daytona in 1967, the Mk IV's became the selected choice for the rest of the 1967 season. They won the Sebring race that year and had the famous 1-2-3 finish at Le Mans. To have an all-American winning car in 1967, having started work from a clean sheet of paper in 1963 was a thrilling culmination to the program.

The FIA changed the rules again for 1968 and limited the engine size to five liters. There was no official Ford factory team in 1968, but John Wyer continued upgrading the original five-liter GT 40 and won Le Mans and other events in 1968 and 69.

The Ford GT became an icon and is now being revived in its own image with the new Ford GT sports car.

Appendix II

Dan Gurney's Tribute Speech

Carroll Shelby was young when both motor racing and the automotive industry were young – barely half the age that they are now. Our country was less exhausted then, exhibiting a youthful freedom that provided the right atmosphere for a man of his talents and ambitions. He was a man of contradictions. Kind on one hand and very cunning on another. Charming and mean. Romantic and rough. A friend to many and a foe to some. He loved cars, women and business. As a salesman, he had no peers. As a negotiator, he would've given Bernie Ecclestone a run for his money. He had charisma and courage. He was a visionary.

As a driver, Carroll was no slouch. I rated him very highly. I think he could've had a successful Formula 1 career if he'd been able to continue.

Carroll suffered very severe health problems for most of his life. He had a heart transplant, a kidney transplant; he inspired and endured the ups and downs of big-time motor racing with its ecstasies and tragedies; he built a car empire; he had multiple marriages; he flew planes and cooked chili; he traveled the world and hunted in Africa. The mind boggles at this man's energy and spirit.

It's a cliché to say they threw away the mold when they made him. In his case, that is a good thing, because whoever would try to make a new one would get sued by him for copyright infringement.

Appendix III

Ford Motor Company

Edsel B. Ford II
Board Director

December 12, 2018

Chris,

Thank you very much for letting me look through your new book. I think it is going to be a real winner, especially for Carroll Shelby fans.

My "yellow" tabs are just fine!

Merry Christmas,

Edsel

One American Road, Dearborn, Michigan 48126-2798 USA

Appendix IV

FORD | DESIGN

From the office of: Moray Callum
Ford Motor Company
21175 Oakwood Blvd
Dearborn, Michigan 48124

17 November 2017

To Chris Theodore,

Congratulations on being the high bidder on the 2004 Shelby Cobra Concept at the Greensboro Auto Auction on 03 November 2017!

I can attest that this is the one-of-a-kind, hand-built concept that originated from the team you led while at Ford Motor Company. This car debuted at the 2004 North America International Auto Show (NAIAS) in Detroit, Michigan. This car then became the 'hero car' in the 2005 motion picture "XXX: State of the Union". You will notice that the car was signed by Carroll Shelby on the passenger-side firewall area, and by your team on the underside of the driveshaft tunnel.

Regards,

Moray Callum

Moray Callum

Vice President, Ford Motor Company - Global Design

APPENDICES

Appendix V

Viper Project Team

Engineering Team

Jack H Atabak
    Engineer – body
Ronald F Bauer
    Vehicle Synthesis – Development
Al Bederka
    Design Manager
Craig S Belmonte
    Engineer - Viper Frame
Stan Bialecki
    Interior trim parts
James R Broske
    Engine Build and Test Coordinator
Charles H Brown III
    Engine Design Engineer
Dave Buchesky
    Suspension and Brake Engineering
Michael N Cipponeri
    Electrical / Electronics Manager
Helen B Cost
    Interior Trim
Joseph J Dapsi
    Body and Trim
Michael A Dazy
    Cooling
John M Donato
    Driveline
Sandy J Emerling
    Body Manager
Alton J Fields, Jr
    Fuel, Exhaust, Cooling, A/C Plumbing
Joseph M Gall
    Engine Design Engineer
Peter M Gladysz
    Chassis Manager
Neil Hannemann
    Vehicle Development Engineer
Herbert J Helbig
    Synthesis Manager
John Hill
    Engineering Specialist
Michael A Kasper
    Engine Calibration
Roy F Macuga
    Interior
David B McGeary
    Fuel, Exhaust, Cooling, A/C Plumbing
Kenneth C Nowak
    Vehicle Synthesis – Development
Randy Olejniczak
    Interior
Stephen C Parman
    Interior
Sharad R Patel
    Materials Engineering
Daniel J Pearlman
    Body Panels
Nippani R Rao
    Materials Engineering – Exterior Body
James Royer
    Director of Viper Engine
James L Sayen
    Program Management
Roy Sjoberg
    Chief Engineer
Jerry L Sobota
    Door Sills and Exhaust Covers
Stanley C Surratt
    Cockpit Module Specialist, Body Manager
Dara Tomczak
    Vehicle Synthesis – Development
Frank R VanWulfen
    Electrical
Willem L Weertman
    Engine Specialist
Alvin V White
    Front of Dash
Richard Winkles
    Engine Calibration/Engine Development

Support team

William F Adams
    Lead Technician
George Baliko
    Engine Designer – Checker
Brett G Bamford
    Designer – Vehicle Package Drawings
Jim Barney
    Designer
Larry W Bell
    Design Co-ordination

Elias Betaya
: Craftsperson

Beth Blake
: Designer

Casey Blake
: Designer

Tiffany Bouey
: Administrative Assistant

Robin L Boyer
: Technician – Engine Group

Murray Bursott
: Designer – Chassis

José Castro
: Craftsperson

Phil Clark
: Designer – Suspension

Bert Condron
: Checker – Engine Design

Craig A Confer
: Designer

Michael Copolla
: Mock Up and Bed Plate Coordinator

Christopher T Damitio
: Designer – Engine group

Glenn F Dimmler
: Dimensional Analysis

Mike Fay
: Procurement

Ira L Fisher
: Vehicle Cost

Lydia Fleming
: Executive Secretary

Robert C Guerity
: Public Relations

Steve Gorski
: Designer

John Gray
: Manufacturing

Linda S Gutmann
: Designer

Bill Harbin
: Design Supervisor – Engine Group

Thomas E Highlen
: Designer – Instrument Panel

Ralph Hoag
: Designer – Chassis

Mike Holmes
: Designer

Brian W Hoxsie
: Body Designer

George Irwin
: Electrical Technician

Paul P Ivan
: Engineering Specialist

Don Jankowski
: Vehicle Build/Vehicle Synthesis

Chris Kaminski
: Illustrator

Bob Kane
: Craftsperson

Skip Karp
: Engineering Specialist

Wallace Kaun
: Designer – Chassis

Galen Kerns
: Engineering Specialist

Peter R Kinsler
: Designer - Engine Group

Frederick Krupic
: Stockroom Operations

Roger Kukola
: Designer

Joseph G Kummer
: Service Engineer - Engine Repair Procedures

Leonard LaBuda
: Designer

Richard J Lapointe
: Designer

John Lenosky
: Designer

Howard Lewis
: Plant Manager

Larry McLeese
: Chassis Development at Arizona Proving Grounds

Tom McVetty
: Engineering Specialist

Jean P Mallebay-vacqueur
: General Manager

James D Mather
: Mechanic/Fabricator

Wesley J Meetze
: Mechanic/Fabricator

Dwayne Merritt
: Designer

Jack Nichols
: Manufacturing

Kevin Nunley
: Designer

Bob Oben
: Designer – Body

APPENDICES

Jim Pawlowski
  Designer – Body
Jeffrey D Petrilli
  Mechanic/Fabricator
Perry Plouff
  Illustrator
Juliana Radin
  Administrative Assistant
Larry Rathgeb
  Chassis/Suspension Specialist
Fred Roberts
  Design Coordinator
Ron Rogalski
  Program Timing Coordinator
Rich Ryan
  Designer – Body
David Swietlik
  Procurement and Supply
Tony Schiavinato
  Designer
Raymond T Schilling
  Technician
Harry Shay
  Designer – Body
Brian E Shea
  Designer
William A Sherwood
  Designer
Bill Smith
  Manufacturing
Piet Smitvangelder
  Stress analysis

Larry Storinsky
  Vehicle Timing
Greg Szareme
  Designer – Body
Andy Teevin
  Structure Designer
Jim Vermeesch
  Designer
Mark V Vermeersch
  Designer
Vincent F Vuichard
  Bill of Materials Specialist/Computer Specialist
Roberta K Waterworth
  CATIA Designer - Engine Group
Jeffrey W Watt
  Manufacturing
David K Wellbaum
  Designer
Wiley Wendell
  Designer – Chassis
David M Wittbrodt
  Designer
Walter L Young
  Craftsperson
Ralph Youngdale
  Designer – Chassis
Robert Zeimis
  Engine Development - Build up & Dyno Testing
Bob F Zeimis, Jr
  Designer
Kenneth L Zalewski
  Designer – Interior

THE LAST SHELBY COBRA

Appendix VI

Ford GT Team

Executive Team
Neil Ressler
Mike Zevalkink
John Coletti
Fred Goodnow

Corporate Design
Camilo Pardo
Cynthia Mallia-McIntosh
Lori Ranger Gonzalez

SVT Program Team
Scott Ahlman
LeAnn Ballard
Paul Bang
Geroge Barber
John Pfeiffer
Mark Carpenter
Bill Clarke
Ellen Collins
Bob Damron
Doug DeLong
Corrine Demarco
Jayson Denchak
Pat Dimattia
Kip Ewing
Dan Fisher
Paul Frison
Brian Gabel
Dale Gage
Paul Genis
Primo Goffi
Mary Gonzalez
Tony Greco
Bob Grenkowitz
Greg Grinner
Herbert Gruber
Ken Gutowski
Mitchell Guzzo
Cathy Hagemeyer
Jamal Hameedi
Neil Hannemann
Kent Harrison
Curt Hill
Annette Huebner
Tom Jones
Scott Keefer
Rick Kirk
Mike Landry
Mike Leese
Mark Lenz
Gene Martindale
Glenn Miller
Gene Nagy
Jerry Naper
John Pfeiffer
Scott Pineo
Benjamin Pohl
Tom Reichenbach
Mark Rowe
Paul Senczak
Jeff Silagy
Gerry Silka
Andy Slankardd
Jamie Sullivan
Mike Sullivan
Scott Tate
Tom Technos
Nick Terzes
Hau Thai-Tang
Brian Tone
Sid Wells
Chris White

Electrical & Electronics Team
Gwen Aid
Randy Betki
Dave Burdette
Terry Cwik
Antonio D'Agostino
Charles Elkins
Fred Hendershot
Rita LaFaive
Bryan Madonna
Don Price
Marty Robinson
John Tenbusch
John Vangelow

Manufacturing & Materials
Adrian Elliott
Rama Koganti
Kim Lazarz
Mathew Zaluzec

APPENDICES

Research Safety Team
Ari Caliskan
Xiaoming Chen
John Jaranson
Richard Jeryan
Tim Keer
Jeff Taylor

Roush Team
Adam Alford
Mike Beaubien
Mark Brasseru
Denis Breitenbach
Rich Brooks
Bob Brown
Rick Brzys
Dan Coccimiglio
Mike Deneau
Dan Duszkiewicz
Eric Ferrell
Nicholas Frania
Beth Friday
Rober Gardner
Jim Gilsbach
Paul Gordanier
Anthony Hallup
Keith Healy
Wade Huber
Gary Jurik
Jeremy Keller
Michael Kramer
Eric Lichtfusz
Tom Logar
Cheryl Machovec
Karsten Mathies
Brian McAllister
Joe Meyers
Ron Muccioli
Dan Paeth
Jason Rem
Paul Richards
Jay Schultz
Louis Sdao
Tom Siebyla
Eric Smith
Wendy Strauss
Mike Sturr
Al Szilagi
Jason Taylor
Mike Terpak
Troy Thompson
Chuck Thornton
Chris Ward
John Wasner
Greg White
Craig Wood
Jim Yagley

THE LAST SHELBY COBRA

Appendix VII

Daisy and GR1 Project Teams

Design, Valencia Studio

J Mays
VP Design

Scott Strong
Advanced Design

Richard Hutting
Daisy Design Chief, Valencia Studio

George Saridakis
GR1 Design, Irvine Studio

Advanced Product Creation

Chris Theodore
VP Advanced Product Creation

Hermann Salenbauch
APC Director

Manfred Rumpel
APC Manager

Mark Bergdahl
Daisy Design Leader

Mac Crossett
Frame design/FEA

Gennadly Goldenshteyn
Driveline/Transaxle

Ed Hosni
Dry sump/rear suspension/fuel system

Andy Lane
Brakes & Pedals/vehicle planning

Nick Rallis
Powertrain & driveline/exhaust

Bill Stinnett
Front suspension/vehicle dynamics/electrical

Barry Alexander
Engine & cooling/digital buck mechanical design

Dave Burgess
Rear suspension & transaxle/electrical design

Art Goudie
Front suspension/steering design

Geroge Turczyn
Manager – NAC CAD chassis & Package

Clarence Smith
Supervisor – NAC CAE chassis & Package

Dave Chapp
Park brake & shifter/engine mount design

Ron Hnatio
Torque tube/clutch/oil system design

Gary Kohlsmith
Exhaust/headers/fuel system design

Dan Lafferty
Frame/people package design

Sean Remisoski
CAD data translation

Advanced Engine

Adam Gryglak
Advanced Engine Manager

Kevin Byrd
V-10 Project Lead

Greg Coleman
Technologist/Engine Build/Dry Sump

Nathan Fogg
Intake/Exhaust System

Mike Shelby
Engine Calibration

Dan Cross
CAD Senior/data translation

Keith Gosfzinski
CAD front engine accessory drive

John Lanzetta
Supervisor – Advanced Transmission

Ed Olin
CAD overall systems/FEAD

Jim O'Neill
Engineering – Powertrain Systems

Chris Wicks
CAD Block/Knee Brace

Rick Williams
Crankshaft/Damper

Judy Elmore
Financial Analyst/CPARS

Research and Advanced Engineering

Gerhard Schmidt
Vice President Research

Graham Hoare
Powertrain

Charles Wu
Manufacturing

Priya Prasad
Safety

Matt Zaluzek
Senior Staff Tech Specialist

Adrian Elliott
Advanced joining & body construction

Richard Jeryan
Staff Tech Specialist – structural materials

APPENDICES

Rich Antoun
    Manager – CVL/Heat management & Aero
Rob Lietz
    St Tech Specialist

Key Suppliers

Metro Technologies, Craig Blust
    Frame Build
Roush Industries, Jay Schultz
    Vehicle Build
Technosports, Bob Nowakowski
    Chassis Build
Auto Tech, Kent Anderson
    Sparco Seats/Interior
Michelin, Dave Handy
    Show Tires
BBS, Mike Kaptuch
    Wheels
Ricardo, Tony Walsh
    Transaxle
Superform
    GR1 Aluminum Body Panels
ZF
    Steering Gear/Pump
Dura
    Pedals/Electric Parking Brake
Aria
    GR1 Body Fabrication and Show Car Assembly

Appendix VIII

Shelby GT500 Condor Project Team

Design

J Mays
 VP Design
Doug Gaffka
 Design Lead

SVT – Special Vehicle Team

John Coletti
 Chief Engineer
Ellen Collins
 Program Manager
Mark Lenz
 Program Management
Celeste Kupczewski
 Program Management
Nick Terzes
 Body Engineer
Scott Tate
 Body Engineer
Joe Buchwitz
 Chassis Engineer
Tim Smith
 Chassis Engineer
Enzo Campagnolo
 Chassis Engineer
Dean Martin
 Chassis Engineer
Brian Roback
 Powertrain Engineer
Ed League
 Powertrain Engineer
Dave Dempster
 Powertrain Engineer

Mustang Team Support

Hau Thai-Tang and the entire S197 Mustang Team

More from Veloce Publishing ...

Ford GT – Then, and Now

This book provides a different view of the Ford GT legend, featuring driver interviews, and both historical and new photographic records. Some of the old myths and legends of the racetrack have been revisited, Ford GT drivers have been given their place on the roll of honour, and their opponents on the racing track are also discussed. With details of the Ford GT replica industry, this book is an historical and interesting record of this classic car, and a must for all GT enthusiasts.

Hardback • 24.6x24.6cm • 240 pages • 491 pictures

ISBN: 978-1-787111-26-4

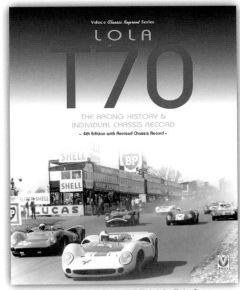

ISBN: 978-1-787110-51-9

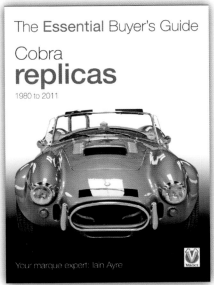

ISBN: 978-1-845843-95-3

For more information and price details, visit our website at www.veloce.co.uk • email: info@veloce.co.uk • Tel: +44(0)1305 260068

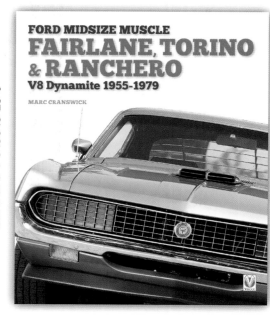

ISBN: 978-1-845849-29-0

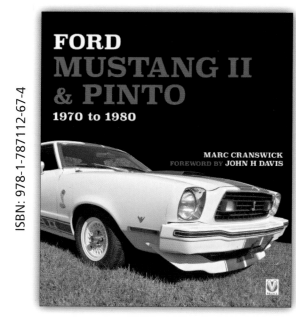

ISBN: 978-1-787112-67-4

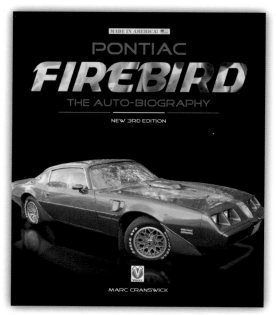

ISBN: 978-1-787110-71-7

ISBN: 978-1-78711-003-8

For more information and price details, visit our website at www.veloce.co.uk • email: info@veloce.co.uk • Tel: +44(0)1305 260068

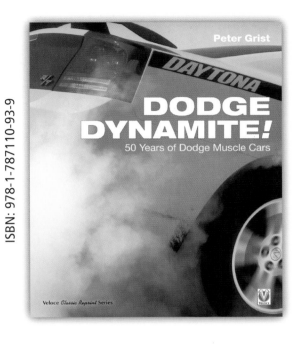

ISBN: 978-1-787110-93-9

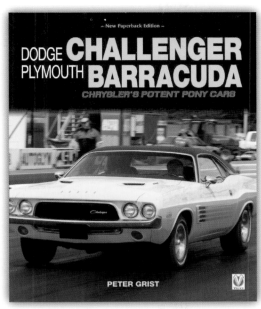

ISBN: 978-1-787110-94-6

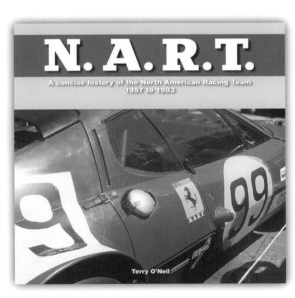

ISBN: 978-1-845847-87-6

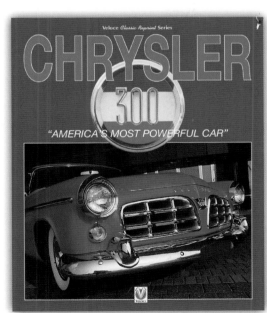

ISBN: 978-1-845849-61-0

For more information and price details, visit our website
at www.veloce.co.uk • email: info@veloce.co.uk • Tel: +44(0)1305 260068

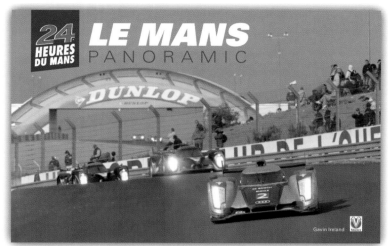

ISBN: 978-1-845842-43-7

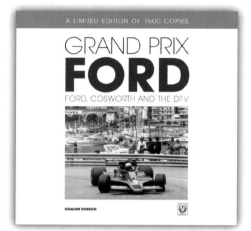

ISBN: 978-1-845846-24-4

ISBN: 978-1-845846-23-7

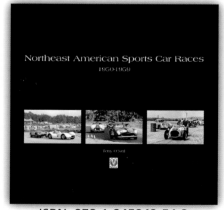

ISBN: 978-1-845842-54-3

For more information and price details, visit our website at www.veloce.co.uk • email: info@veloce.co.uk • Tel: +44(0)1305 260068

INDEX

AC Ace 103
Allen, Tim 23
AMC (American Motors) 17, 18, 20, 27, 37, 83, 101
Amelia Island Concours 8, 101, 130-133
Anderson, Pamela 73
Andretti, Mario 14, 73
Aptera 92
Aria 67, 153
Arnold, Klaus 97
Aston Martin 10, 14, 115, 138, 139
Autokraft Mark IV 18
Automotive Hall of Fame 25, 38
Automotive News 29, 118
AutoWeek 58, 111, 118

Bailey, Len 140
Barrett-Jackson 74, 81-83, 85, 88, 91, 102, 107, 109, 123
Basalt 94
Baylor Hospital 117
Bedore, Dan 53
Bel-Air Country Club 42, 60, 85
Belasco Theater 116
Bell, Kenny 109
Bez, Ulrich 8, 115
Bizzarini 27
BMW 69, 72, 77
Boeckman, Beau 131
Boeckman, Bert 92
Bolyn, Amy 78, 83, 84, 90
Bondurant, Bob 29, 32
Borda, Walter 124
Bordinat, Gene 139
Borg-Warner 22
Boyd, Gordon 34
Brabham, Jack 10
Bridgehampton Race Track 13
Broadley, Eric 38, 139, 140
Brock, Pete 26, 60, 62, 96, 99, 100, 111, 134
Brown, David 10
Brutsman, Bud 8, 9, 43-47, 49, 51-53, 55, 57
Bryant, Thomas L 58
Bush, George HW 77
Byrd, Kevin 45, 152
Byron, Dave 8, 91

Cadillac 114
Callum, Moray 129-131, 146
Campana, Joe 19
CAR 62
Car and Driver 14, 29
Carducci, Lee 20
Carroll Shelby Foundation 74, 81, 101, 103, 107
Cars & Concepts 77
Castaing, François 6, 17-19, 25, 26, 83, 101
Chaparral 14
Chapman, Colin 73, 83, 96, 134, 139

Chaunac, Hughes de 26
Chevrolet
   Corvette 12, 40, 45, 50, 55, 90, 96, 109, 113, 114, 124
   SSR 76, 91
Chinetti, Luigi 131
Christie's auction 40, 42
Chrysler Corporation 5, 6, 8, 9, 15-23, 25-27
Citroën 83
Clark, Jim 101
Clinard, John 8, 29, 45, 104, 130
Cluxton, Harley 9, 82, 84, 85
CNW Market Research 73
Coleman, Greg 45-47, 50, 152
Coletti, John 8, 9, 29-36, 40, 42, 43, 45, 68, 71, 73, 97, 104
Concours d'Elegance of America 132, 134
Condor 27, 71, 73, 154
Conway, Joe 104, 114, 121
Cooper 99, 139
Cosby, Bill 81
Crain, Keith 8, 29, 98, 104
Cube, Ice 117

Daimler-Benz 26, 77
DaimlerChrysler 26, 27, 109
Daisy (Shelby Cobra Concept) 5, 8, 42-60, 62, 66, 67, 81, 92, 94, 103, 113, 114, 117, 118, 124-135, 152
Davis, David E 104
Davis, Gary 120
Davis, Steve 85
Daytona Speedway 26, 141-143
Decker, Dave 127
DeLorean 14, 16
DenBeste, Bill 114
Dennis, Ron 20
Detroit Athletic Club 58
Detroit Free Press 55, 57
Detroit News 55
Dingman, Michael 109
Dodge
   Dakota, Lancer, Shadow, Omni, Omni GLH 16
   Ram 18, 19
   Shelby Charger 16
   Spirit R/T 21
   Stealth 21
   Viper 5, 8, 17, 19-27, 33, 34
Dodge, Horace 29
Donohue, Mark 12
Duntov, Zora 97
Duprey, Gene 8, 132

Eaton, Bob 26
Economaki, Chris 10
Edelbrock 84
Edgar, William 95
Edsall, Larry 8, 9, 28, 35, 127
Erickson, Larry 72

Fair Lane 127-130
Farley, Jim 8, 98, 103, 115, 116, 121
FCA (Fiat Chrysler Automobiles) 9, 21, 25
Fernandez, John 8, 9, 16-18, 25, 26, 104
Ferrari automobiles 12, 13, 19, 20, 26, 40, 45, 54, 85, 138, 139, 140
Ferrari, Enzo 13-15, 96, 98, 120
Fiat 16, 31
Fields, Al 16
Fields, Mark 77, 78, 79, 105, 114, 117
Firestone 28
Fisker, Henrik 62
Fittipaldi, Emerson 131
Foose, Chip 103
Ford Vehicles
   Bronco 43
   Crown Victoria 78
   Expedition 78
   Explorer 28
   F-150 34, 91, 123
   Fairmont 28, 69
   Focus 28, 29
   Ford GT (GT40 and Petunia) 5, 8, 13, 14, 26, 27, 28, 30, 31, 35, 37-40, 42-45, 47, 50, 53-55, 58, 60, 62, 65, 67, 68, 72, 73, 77, 78, 80, 81, 86-89, 92, 94, 96, 97, 108, 110, 114, 118, 120, 123-125, 127, 135, 138, 140, 143
   Forty-Nine 33
   Mustang 5, 8, 12, 15, 28-31, 44, 45, 52, 54, 55, 58, 68, 69-74, 77-79, 81, 90, 91, 95, 102-105, 107, 118, 121, 123, 138, 139, 154
   Ranger 78
   Sport Trac 78
   Taurus 28
   Thunderbird 69, 77, 78, 97
Ford Performance Technical Center 123
Ford, Bill Jr 27, 32, 33, 35, 40, 54, 55, 77
Ford, Clara 127
Ford, Edsel 27
Ford, Edsel II 8, 16, 27-30, 34, 73, 90, 97, 98, 104, 114-116, 120, 129, 130, 145
Ford, Henry 14, 26, 124, 127-129
Ford, Henry II 16, 28, 30, 32, 91, 96
Forghieri, Mauro 19, 20
Foyt, A J 10, 14, 30, 97
Frankenheimer, John 33
Frey, Don 13, 45, 138, 139
Friedkin, Tom 96, 104

GAA Classic Cars 127
Gaffka, Doug 69, 73, 154
Gale, Tom 19, 26

Gall, Joe 8, 147
Galpin Ford 92
Garner, James 83, 101
Geddes, Ray 139
Gilles, Ralph 131
Gladysz, Pete 16
GM (General Motors) 30, 77, 78, 96, 97, 101, 114, 125, 131
Goodnow, Fred 8, 27, 31, 33, 34, 40, 41, 80, 150
Goodwood Festival of Speed 40
Goodyear 78, 85
Grady, Jim 22
Grand Prix 32
Graziano 113
Green, Dean 128
Greenfield Village 30
Guindon, Dick 57
Gurney, Dan 11, 12, 13, 29-31, 36, 40-42, 62, 91, 92, 97, 120, 136, 144
Gurney, Evi 8, 29, 120

Hall, Jim 12
Hameedi, Jamal 33, 85, 150
Hannemann, Neil 8, 14-17, 25, 26, 33-35, 38, 41, 104, 147, 150
Henry, Charlie 16
Henry Ford Estate 127-129
Hermann, Brian 8, 132
Hertz 78, 79
Hill, Graham 10
Hill, Phil 10, 12, 26, 29, 32
Hilton, Baron 95, 104
Holman-Moody 37, 143
Hoover, Bob 94, 120
Howard, Benny 98
HST Automotive 110
Hubbach, Bob 26
Hutting, Richard 43, 44, 47-50, 53, 70, 152

Iacocca, Lee 15-19, 23, 28, 29, 45, 73, 96, 138-140
Indy 500 (Indianapolis) 4, 10, 22, 23, 25, 26, 69, 139, 140
Irvine Studio 43, 60, 62, 67, 152
Irwindale Speedway 52, 53, 55, 62, 128, 132
Ito, Kunihisa 31

Jacuzzi, Joe 78
Jaguar 13, 69
Jankowski, Don 16
Jeanes, William 8, 28, 29, 120
Jeep 5, 17, 18, 37, 101
Jennings, Jean 104
Johnson, Tony 33, 34
Jones, Parnelli 10, 29

Kar-Kraft 141
Keebler, Jack 19
Kent, John 18

INDEX

Klein, Calvin 73
Kobe/Wagyu Beef 95, 96
Kowaleski, Tom 21, 23, 26
Kuzak, Derrick 77, 78, 80, 81, 86

Laguna Seca Raceway 11, 40, 41
Lajdziak, Jill 125
Lamborghini 14, 17, 19, 20, 26
Las Vegas Speedway 30, 40
Lauren, Ralph 73
Laviolette, Vince 109, 134
Lawrence Technological University 124
Le Mans 10, 13, 14, 26-28, 30, 31, 33, 34, 42, 54, 60, 62, 78, 96, 97, 115, 120, 138-143
Lear 34
Ledwinka, Hans 83
Legend Industries 16
Leno, Jay 14, 23, 24, 40, 42, 98, 120
Lieberman, Jonny 109
Limongelli, Joey 8
Lincoln 42, 69, 71, 77, 78
Lola 138, 139, 140
Lotus 14, 21, 83, 94, 96, 139
Luft, John 93, 121
Lunn, Roy 17, 37, 38, 138
Lutz, Bob 6, 8, 18-22, 25, 26

Manthey, Steven Charles 92
Markus, Frank 8, 110
Martindale, Gene 114, 150
Maserati 11, 95, 115, 138
Mason, George 17
Mayflower 33
Mays, J 27-29, 31, 33, 43-45, 47, 49, 51, 53, 54, 56, 58, 66, 67, 69, 70, 86, 88
Mazda 47, 95
McCluggage, Denise 97, 104, 120
McQueen, Chad 30
McQueen, Steve 32
McGraw, Jim 120
McKenzie, Angus 109
McLaren 12, 20, 140
Mees, Hubert 94
Meguiar, Barry 104, 117, 131-133
Metro Technologies 50, 57, 153
Miles, Ken 21, 29, 37, 38, 53, 85, 86, 95-97
Miller, Walter 104
Monza Eni Circuit 13
Moss, Stirling 11, 12, 29
Motor Trend 8-11, 17, 22, 28, 36, 41, 51, 53, 58, 59, 74, 75, 81, 82, 84, 94, 106, 109, 111, 114, 118, 121, 135
Mountain, Chuck 140
Mulally, Alan 105
Mullins, Kathleen 128, 129
Multimatic 30
Musk, Elon 37, 94

Nagara, Keith 124
NAIAS (North American International Auto Show) 19, 31, 54, 58, 67, 68
Najjar, Johnny 139
Nardo Ring 40
Nash/Kelvinator 17, 101
Nasser, Jacques 26-31, 33
Neale, Bill 103, 120
Negsted, Bob 97, 110
Newey, Adrian 83
New York Times 11, 118
Nowak, Ken 16
Nowakowski, Bob 8, 123, 129
Nowakowski, Mike (Mikey) 8, 129, 132
Nürburgring 26

O'Connor, Jim 35
Oldsmobile 30, 97
Olson, Warren 96
Oreca 26
OX2 92, 93, 94

Paradise, Phil 138
Pardo, Camilo 27, 60, 62
Parry Jones, Richard 27, 31, 62
Partial Forced Induction 124
Passino, Jacque 13
Patterson, Gary 8, 40, 78, 110, 117, 120, 121, 123
Paxton/Vortech 84, 90
Pebble Beach Concours 26, 29, 35, 37, 62, 66, 115
PEMF (Pulsed Electromagnetic Field) 95
Penske, Roger 14, 125-127, 134
Petersen Museum 90, 118, 120
Petersen, Don 138
Petty, Richard 73
Petunia 8, 27, 28, 31, 43, 45
Piech, Ferdinand 27
Pininfarina, Andrea 31, 32
Plymouth Prowler 33
Porsche 14, 43, 44, 54, 120
Prasad, Priya 124
Pratte, Ron 85
Prechter, Heinz 29, 77
Precision Race Service 127
Price, Bud 57
Priestley, Jason 43
Prudhomme, Don 102
Purdy, Ken 10, 22

Quadricycle 124, 125

Rager, Don 31, 97
Rayhal, Bobby 14, 69
Redman, Brian 26
Reichenbach, Tom 34, 38, 150
Remington, Phil 21, 62, 96, 140
Renault 17, 37
Reno Championship Air Races 95
Ressler, Neil 8, 27, 30, 31, 33, 34, 40, 42, 81, 150
Reuss, Mark 131
Reuter, Rick 101
Reventlow, Lance 96
Rewey, Bob 26
Rheims 26
Ricardo 86, 153
Rides 5, 43, 45, 52, 53, 57, 58, 60, 112, 130, 137
Riverside International Motor raceway 11, 95, 96, 142
Road & Track 9, 10, 14, 15, 19, 58, 84
Rock, John 30, 97
Rod & Custom 10
Rodriguez, Pedro 12
Romney, George 17
Roush 19, 33, 35
Rover 69
Royer, Jim 20, 147
Ruby, Lloyd 32
Rumpel, Manfred 8, 43-45, 62, 72, 85, 129, 130, 152
Rushwin, Shamel 26, 27, 109, 124

Saarinen, Eero 33
SAE (Society of Automotive Engineers) 37, 109, 110, 138, 142
Safir GT40 Spares Ltd 87
Saleen 33-35, 91, 92, 99, 117, 125
Salenbauch, Hermann 8, 44, 47, 60, 68, 69, 86, 152
Salvadori, Roy 11, 97
Saridakis, George 60, 61, 66, 105, 152

Saturn 125
Sayen, Jim 26
Scheele, Nick 32, 33, 35, 60, 71
Schwartzkopf, Norman 22
Sebring International Raceway 11, 114, 141-143
SEMA 16, 30, 86, 102, 114-116
Shabes, Lenny 112, 117
Shelby American 5, 8, 9, 23, 30, 34, 40, 49, 73, 76, 77, 79, 81, 82, 84, 89-91, 93, 97, 100, 105, 107, 110-114, 116-118, 121, 123, 137, 141
Shelby Automobiles Inc 16
Shelby Automotive Technology Center 120
Shelby International 121
Shelby Performance Center 16
Shelby Vehicles
  1000 (Mustang) 121
  Baha Raptor 123
  Can-Am Spec Racer 16, 17
  Charger 16
  Cobra 5, 12, 13, 15, 18, 20, 21, 23, 26, 28, 30, 35, 43, 45, 47, 49, 51, 53-55, 58, 60, 62, 65, 67, 81-83, 85, 86, 90, 91, 94, 96-99, 101-103, 109-120, 122, 124, 125, 127-129, 134, 135, 137, 139-141, 146
  Cobra Concept (Daisy) 5, 8, 42-60, 62, 66, 67, 81, 92, 94, 103, 113, 114, 117, 118, 124-135, 152
  Cobra Super Snake 8, 81, 84, 85, 99
  Cobra Super Snake II (C/S Series II) 8, 84-86, 89, 92, 94, 104, 109-112, 114, 122-124, 127, 130, 136
  Daytona 13, 14, 25, 26, 55, 60, 62, 81, 97, 99, 100, 120
  GLH-S 16
  GR1 8, 60-68, 81, 83, 105, 113, 127, 136, 152, 153
  GT (MkII and MkIV) 76, 78-81, 83, 136
  GT (Mustang) and GT-H 79, 81, 82, 90, 91, 121, 123
  GT/SC 121
  GT350 12, 13, 28, 54, 55, 79, 81, 96, 101, 103, 106, 109, 114, 115
  GT500 8, 12, 28, 70, 72-75, 81, 82, 86, 88, 89-91, 105, 107, 109, 114, 116, 117, 121, 154
  GTS 121
  King Cobra (Cooper Ford) 120
  KR (King of the Road) 28, 88-91, 105
  Super Snake (Mustang) 89, 90, 109, 121, 122
  Terlingua 89, 90
Shelby, Aaron 5, 8, 9, 23, 85, 104, 114, 123
Shelby, Cleo 29, 35, 57, 58, 66, 85, 95, 97, 101, 102, 116, 117
Shelby, Lena 25
Shelby, Michael 95
Shelby, Pat 104
Shelby, Patrick 104
Shelby, Sharon 104
Shilling, Pat 16
Sjoberg, Roy 8, 20-22, 26
Smith, Leah 120
Spa 26
Sparco 51, 54, 111, 124, 153
Spinella, Art 73
Spirit of Ford Award 29, 30, 97
Special Projects 129
Sperlich, Hal 139

Sports Car Graphic 10
Sports Illustrated 11
Stallkamp, Tom 19
Stewart, Jackie 14, 29, 32, 35, 36, 40
Stone, Matt 53, 58, 85, 104, 128
Stoner, Greg 128, 129
Superform 67, 152
SVT 27, 30, 33, 34, 43, 45, 71-74, 85, 86, 90, 91, 150, 154
SVT Vehicles
  Condor 27, 71, 73, 154
  Focus 29
  Lightning 45, 72
  Terminator 27, 71
  Tomcat 72

Teague, Richard 27
Technosports 8, 50, 57, 85, 110, 114, 123, 125, 127, 129, 131, 153
Tesla 94
The Last American CEO 8
Theodore, Tracee 134, 146, 152
Thyseen-Krupp 89
TLC/Discovery 43, 57, 58, 62, 112
Toyota 95, 98, 99
Tronto, Alfredo Dr 95
Tucker, Preston 14
Turner, Al 19

UCLA Medical Center 117
Ungemach, Doug 78
Uni-Chassis 83, 85, 86, 89, 109-111, 113, 124, 125, 134
Unique Performance 102
USA Today 118

V-10 18-22, 43-47, 49, 50, 67, 127, 128, 130, 131
Valencia Studio 43, 49, 50, 57, 70, 132, 152
Van Tune, C 22, 23
Venture Industries 97
Vines, Betsy 127, 129
Vines, Jason 8, 29, 127
Volvo 97, 109
VW 27

Walker Rob 10
Walker, Howard 8, 131
Wall Street Journal 27, 118
Wallace, Tom 124
Ward, Rodger 10
Ward's AutoWorld 92
Warner, Bill 8, 9, 11, 98, 99, 101, 104, 130, 131
Way, Kenneth 34
What did Jesus Drive? 8
Wilbur, Paul 77, 80
Willow Springs Raceway 16, 23, 54
Winjet, Larry 30, 97, 114
Winkles, Dick 8, 16, 17, 21-24
Wixom Assembly Plant 39, 42, 77
Wood, Larry 109
Woodward Avenue 10, 33
Wright, Frank Lloyd 136
Wyer, John 37, 139, 143

xXx: State of the Union 117, 127

Yanez, Ken 129
Yates, Brock 6, 104
Yeager, Chuck 95

Zarella, Ron 30
Zemis, Bob 20
Zevalkink, Mike 8, 30, 33, 37, 40, 42, 68, 77